BIG DATA DEVELOPMENT

大数据开发实战

猿媛之家／组编

韦宇杰 楚秦 等／编著

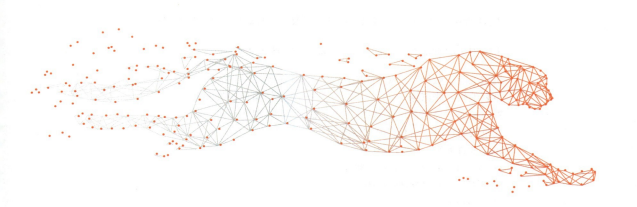

本书深入探讨了大数据技术的核心概念和实际应用。从大数据的基础架构 Hadoop 开始，逐步解析了分布式协调服务 Zookeeper、数据仓库 Hive、面向列的数据库 HBase 等关键技术。此外，还介绍了数据迁移工具 Sqoop、数据采集工具 Flume、发布订阅消息系统 Kafka 等实用工具。本书还深入讲解了数据处理分析引擎 Spark、全文搜索引擎 Elasticsearch 及分布式处理引擎 Flink 的工作原理和应用实例。最后，通过电商推荐系统实战和 Flink 实现电商用户行为分析两个案例，展示了大数据技术在实际业务中的应用。

本书附带全书实例源代码、电子版本教程（下载方式见封底），以及可扫码观看的长达 13 个小时的部分实例操作视频，帮助读者更深入了解大数据技术的具体内容，非常适合对大数据技术感兴趣的读者，尤其是想要深入了解大数据技术原理和应用的开发者和管理者阅读。

图书在版编目（CIP）数据

大数据开发实战／猿媛之家组编；韦宇杰等编著. —北京：机械工业出版社，2024.7

ISBN 978-7-111-75693-4

Ⅰ.①大…　Ⅱ.①猿…　②韦…　Ⅲ.①数据处理软件　Ⅳ.①TP274

中国国家版本馆 CIP 数据核字（2024）第 088822 号

机械工业出版社（北京市百万庄大街 22 号　邮政编码 100037）
策划编辑：张淑谦　　　　　　　　　　　责任编辑：张淑谦
责任校对：高凯月　张慧敏　景　飞　　　责任印制：邓　博
北京盛通数码印刷有限公司印刷
2024 年 11 月第 1 版第 1 次印刷
184mm×240mm · 19.5 印张 · 488 千字
标准书号：ISBN 978-7-111-75693-4
定价：119.00 元

电话服务　　　　　　　　　　　网络服务
客服电话：010-88361066　　　　机　工　官　网：www.cmpbook.com
　　　　　010-88379833　　　　机　工　官　博：weibo.com/cmp1952
　　　　　010-68326294　　　　金　书　网：www.golden-book.com
封底无防伪标均为盗版　　　　　机工教育服务网：www.cmpedu.com

前 言
PREFACE

大数据是未来的发展趋势,将应用到生活各个方面,影响和改变我们的生活。本书旨在使更多读者更深入地了解大数据,以及使用大数据的技术解决生活中的问题。

本书采用基础优先的方法,每章内容都有使用大数据组件安装的实战案例,以及每一个组件的开发案例,并且提供了相应的代码。使用这种理论结合实践的方式,将大数据的技术应用到项目中,从而让读者能更加容易地掌握相关的技术和原理。

本书内容根据笔者在大数据开发中积累的常用技术和经验编写而成,把书中的内容串联在一起,就是日常大数据开发的整个流程,本书先介绍理论,后经过案例练习,注重实战,让读者在阅读的过程中,就能感受到大数据日常开发的一个具体流程,同时也能帮助读者学会大数据项目开发的流程。

本书还介绍了推荐项目的开发流程,详细介绍了项目的需求和实现思路,同时提供了相应的实例代码,供读者学习和参考。书中的推荐项目能够让读者更加深刻地了解大数据在日常生活中的应用,以及如何把大数据的技术应用到项目中。

新手、基础比较薄弱和知识系统不够全面的读者,可以结合本书中的案例,逐一练习。本书配有电子版教程和视频教程,每章的内容都会详细讲解。

本书各章简述部分均列出内容大纲,有助于读者对本章的内容有一个总体的认识和了解,有利于读者以从整体到局部的方式学习各章的内容,达成学习的目标。本书以容易理解的方式讲述各章的内容,教授学习的方法、解决问题的方法、程序设计的概念、代码实现的思路。本书各章内容使用多个简单的例子来演示重要的概念,让读者从实战中领悟到重要概念,化抽象为具体,便于理解和记忆。

本书旨在为读者提供更加简单的方法,让你拥有一本可以实战的书,一本可以带去上班参考使用的大数据技术手册。在阅读过程中有任何问题,欢迎发邮件到 yuancoder@foxmail.com。

环境搭建视频教程二维码清单

课前 01 课前环境视频大纲		课前 02 虚拟机安装	
课前 03 虚拟机安装 Centos7		课前 04 Linux 系统配置	
课前 05 Linux 系统免密登录		课前 06 Linux 系统安装 JDK	
课前 07 Windows 系统安装 JDK		课前 08 Linux 时钟同步配置	
课前 09 IDEA 基本设置			

目录 CONTENTS

前　言
环境搭建视频教程二维码清单

第1章　大数据简介 / 1

1.1　大数据的概述 / 1
1.2　大数据的特点 / 1
1.3　大数据的应用领域 / 2
1.4　目前企业应用的主流大数据技术 / 2
1.5　大数据开发流程 / 3

第2章　大数据基础架构 Hadoop / 5

2.1　Hadoop 简介 / 5
2.2　Hadoop 架构详解 / 5
　2.2.1　分布式存储系统 HDFS / 6
　2.2.2　分布式资源管理框架 YARN / 7
　2.2.3　离线计算框架 MapReduce / 8
2.3　HDFS 读、写文件流程 / 10
　2.3.1　HDFS 写文件流程 / 10
　2.3.2　HDFS 读文件流程 / 11
2.4　HDFS 的实战操作 / 12
　2.4.1　HDFS 上传文件 / 12
　2.4.2　HDFS 创建文件 / 13
　2.4.3　HDFS 创建目录 / 13
　2.4.4　HDFS 重命名文件 / 14
　2.4.5　HDFS 删除文件 / 14
2.5　YARN 原理 / 15
2.6　YARN 调度器详解 / 16
2.7　MapReduce 工作原理 / 18
2.8　MapReduce 核心的原理 Shuffle / 20
　2.8.1　Map 端 / 21
　2.8.2　Reduce 端 / 21
2.9　MapReduce 常用三大组件 / 22
　2.9.1　MapReduce 中的 Partitioner / 22
　2.9.2　MapReduce 中的 Sort / 25
　2.9.3　MapReduce 中的 Combiner / 28
2.10　MapReduce 项目实战 / 29
　2.10.1　清洗日志 / 29
　2.10.2　统计电影最高评分 / 30

第3章　分布式协调服务 Zookeeper / 33

3.1　ZooKeeper 简介 / 33
3.2　ZooKeeper 结构和工作原理 / 33
　3.2.1　ZooKeeper 集群角色 / 34
　3.2.2　ZooKeeper 的数据结构 / 34
　3.2.3　ZooKeeper 的工作流程 / 35
　3.2.4　ZooKeeper 的监听器 / 36
3.3　ZooKeeper 实战 / 38
　3.3.1　ZooKeeper 创建持久节点 / 38
　3.3.2　ZooKeeper 创建临时节点 / 39
　3.3.3　ZooKeeper 递归创建节点 / 39
　3.3.4　ZooKeeper 读取数据 / 40
　3.3.5　ZooKeeper 更新数据 / 41

3.3.6　ZooKeeper 监听节点　/　41
3.3.7　ZooKeeper 监听子节点　/　43
3.3.8　ZooKeeper 实现服务注册与发现　/　44

第4章　数据仓库 Hive　/　49

4.1　Hive 简介和特点　/　49
4.2　Hive 结构和原理　/　50
 4.2.1　Hive 结构　/　50
 4.2.2　Hive 运行的流程　/　51
 4.2.3　Hive 的 HQL 转换过程　/　52
4.3　电商用户行为分析　/　52
 4.3.1　项目背景及目的　/　52
 4.3.2　数据导入　/　53
 4.3.3　数据清洗　/　54
 4.3.4　数据分析　/　54

第5章　面向列的数据库 HBase　/　57

5.1　HBase 简介　/　57
5.2　HBase 架构　/　57
 5.2.1　HBase 的组件　/　58
 5.2.2　HBase 工作机制　/　59
5.3　HBase 数据模型　/　59
5.4　HBase 读写流程　/　60
 5.4.1　HBase 写操作流程　/　60
 5.4.2　HBase 读操作流程　/　61
5.5　HBase 的 API 示例　/　62
 5.5.1　HBase 创建表　/　62
 5.5.2　HBase 保持数据　/　63
 5.5.3　HBase 更新数据　/　64
 5.5.4　HBase 获取数据　/　65
 5.5.5　HBase 删除数据　/　66

 5.5.6　使用 HBase 获取某一行数据　/　67
5.6　HBase 存储订单案例　/　68

第6章　数据迁移工具 Sqoop　/　74

6.1　Sqoop 架构和工作原理　/　74
 6.1.1　Sqoop 导入原理　/　75
 6.1.2　Sqoop 导出原理　/　76
6.2　Sqoop 将 HDFS 数据导入 MySQL　/　77
6.3　Sqoop 将 MySQL 数据导入 HDFS　/　80

第7章　数据采集工具 Flume　/　83

7.1　Flume 简介　/　83
7.2　Flume 构成和工作原理　/　84
 7.2.1　Flume 构成　/　84
 7.2.2　Flume 工作原理　/　84
7.3　Flume 实战　/　85
 7.3.1　Flume 监听目录实战　/　85
 7.3.2　Flume 一对多实战　/　86
 7.3.3　Flume 拦截器实战　/　88
 7.3.4　Flume 采集数据到 HDFS　/　90
 7.3.5　Kafka 对接 Flume 实战　/　91

第8章　发布订阅消息系统 Kafka　/　93

8.1　Kafka 简介　/　93
8.2　Kafka 的消息生产者　/　95
 8.2.1　Kafka 生产者的运行流程　/　95
 8.2.2　Kafka 生产者分区　/　96
 8.2.3　副本的同步复制和异步复制　/　97
 8.2.4　Kafka 消息发送确认机制　/　98
8.3　Kafka 的 Broker 保存消息　/　99

8.3.1 存储方式与策略 / 99
8.3.2 Topic 创建与删除 / 99
8.4 Kafka 的消息消费者 / 99
8.4.1 消费机制 / 100
8.4.2 消费者组 / 101
8.5 Kafka 的存储机制 / 101
8.5.1 Kafka 主题 Topic / 102
8.5.2 Kafka 分片 Partition / 103
8.5.3 Kafka 日志 Segment File / 104
8.6 Kafka 实战 / 104
8.6.1 Kafka 发送消息 / 104
8.6.2 Kafka 自定义分区发送消息 / 106
8.6.3 Spring Boot 整合 Kafka 发送消息 / 108

第 9 章 数据处理分析引擎 Spark / 114

9.1 Spark 简介 / 114
9.2 Spark 运行原理 / 115
9.2.1 Spark 的基本概念 / 115
9.2.2 Spark 运行的原理 / 116
9.2.3 Driver 运行在 Client / 117
9.2.4 Driver 运行在 Worker 节点 / 118
9.3 Spark 算子 RDD / 120
9.3.1 RDD 的属性 / 120
9.3.2 RDD 的依赖关系 / 120
9.3.3 RDD 的 shuffle 过程 / 121
9.3.4 RDD 的缓存和检查机制 / 122
9.4 Spark SQL / 123
9.4.1 Spark SQL 概念 / 123
9.4.2 Spark SQL 的架构 / 124
9.4.3 DataSets 和 DataFrames / 125
9.4.4 Spark SQL 示例 / 126
9.5 Spark Streaming / 127
9.5.1 Spark Streaming 介绍 / 127
9.5.2 DStream 转换操作 / 129
9.5.3 Spark Streaming 窗口操作 / 129
9.5.4 DStream 输入 / 131
9.5.5 DStream 输出 / 134
9.5.6 DSFrame 和 SQL 操作 / 136
9.5.7 Spark Streaming 检查点 / 137
9.6 Spark Streaming 接收 Flume 数据实战 / 137
9.7 Spark Streaming 接收 Kafka 数据实战 / 140

第 10 章 全文搜索引擎 Elasticsearch / 144

10.1 Elasticsearch 简介 / 144
10.2 Elasticsearch 架构和原理 / 145
10.2.1 Elasticsearch 核心概念 / 146
10.2.2 Elasticsearch 工作原理 / 147
10.2.3 Elasticsearch 倒排索引 / 149
10.3 Elasticsearch 实战 / 150
10.3.1 Elasticsearch 索引创建 / 150
10.3.2 Elasticsearch 索引更新 / 151
10.3.3 Elasticsearch 索引查询 / 152
10.3.4 Elasticsearch 索引删除 / 152
10.3.5 Elasticsearch 保存文档 / 153
10.3.6 Elasticsearch 更新文档 / 154
10.3.7 Elasticsearch 精确查询 / 155
10.3.8 Elasticsearch 模糊查询 / 156
10.3.9 Elasticsearch 范围查询 / 157
10.3.10 Elasticsearch 布尔查询 / 158
10.3.11 Elasticsearch 聚合查询 / 159
10.3.12 Elasticsearch 高亮查询 / 160
10.4 Elasticsearch 实现搜索系统 / 161
10.4.1 搜索系统项目环境准备 / 161
10.4.2 Elasticsearch 实现搜索功能 / 167

第11章 分布式处理引擎 Flink / 172

- 11.1 Flink 概述 / 172
- 11.2 Flink 基本组件和运行时架构 / 173
  - 11.2.1 Flink 运行时架构 / 173
  - 11.2.2 Flink 的分层 / 174
- 11.3 Flink 流处理流程 / 175
  - 11.3.1 Flink 环境设置（Environment） / 175
  - 11.3.2 Flink 源算子（Source） / 178
  - 11.3.3 Flink 支持的数据类型 / 179
  - 11.3.4 Flink 转换算子（Transform） / 180
  - 11.3.5 Flink 输出算子（Sink） / 180
- 11.4 Flink 窗口、时间和水位线 / 186
  - 11.4.1 Flink 窗口（Window） / 186
  - 11.4.2 Flink 时间（Time） / 188
  - 11.4.3 Flink 水位线（Watermark） / 191
- 11.5 Flink 状态管理 / 191
  - 11.5.1 Flink 状态分类 / 191
  - 11.5.2 Flink 状态后端 / 191
  - 11.5.3 Flink 状态管理案例 / 192
- 11.6 Flink 处理函数 / 193
  - 11.6.1 Flink 处理函数 / 193
  - 11.6.2 Flink 侧输出流 / 195
- 11.7 Flink 容错机制 / 197
  - 11.7.1 Flink 容错机制概要 / 197
  - 11.7.2 Flink 状态一致性 / 197
  - 11.7.3 Flink 容错机制实战 / 197
- 11.8 Flink 表和 SQL / 198
  - 11.8.1 Flink 表概述 / 198
  - 11.8.2 Flink 结构化函数 / 199
  - 11.8.3 Flink 的 SQL 操作 / 199
- 11.9 Flink 复杂事件处理 / 201
  - 11.9.1 Flink CEP 简介 / 201
  - 11.9.2 Flink CEP 个体模式 / 202
  - 11.9.3 Flink CEP 组合模式 / 204
  - 11.9.4 Flink CEP 超时事件提取 / 206

第12章 电商推荐系统实战 / 208

- 12.1 推荐系统的概述 / 208
  - 12.1.1 推荐系统算法 / 209
  - 12.1.2 推荐系统的数据 / 210
- 12.2 电商推荐系统架构 / 211
  - 12.2.1 电商推荐系统模块 / 211
  - 12.2.2 创建一个推荐项目 / 212
- 12.3 加载数据模块 / 223
  - 12.3.1 模块介绍 / 223
  - 12.3.2 模块实现 / 223
- 12.4 离线统计模块 / 227
  - 12.4.1 模块介绍 / 227
  - 12.4.2 模块实现 / 227
- 12.5 离线推荐模块 / 231
  - 12.5.1 模块介绍 / 231
  - 12.5.2 模块实现 / 231
  - 12.5.3 离线推荐模型评估 / 237
- 12.6 在线推荐模块 / 244
  - 12.6.1 模块介绍 / 244
  - 12.6.2 需求分析 / 245
  - 12.6.3 模块实现 / 247

第13章 Flink 实现电商用户行为分析 / 257

- 13.1 电商用户行为实时分析系统概述 / 257
- 13.2 电商用户行为分析系统架构 / 258
  - 13.2.1 电商用户行为分析系统模块介绍 / 258
  - 13.2.2 创建电商用户行为分析项目 / 259
- 13.3 实时热门商品统计 / 262
  - 13.3.1 实时热门商品统计模块介绍 / 262

13.3.2　实时热门商品统计模块实现　/　262

13.4　实时流量统计　/　268

13.4.1　实时热门页面统计介绍　/　268

13.4.2　实时热门页面统计模块实现　/　268

13.4.3　实时每小时访问量统计模块介绍　/　274

13.4.4　实时每小时访问量统计模块实现　/　274

13.4.5　实时用户访问量统计模块介绍　/　276

13.4.6　实时用户访问量统计模块实现　/　276

13.5　市场营销分析　/　278

13.5.1　市场营销分析——市场推广统计模块介绍　/　278

13.5.2　市场营销分析——市场推广统计模块实现　/　278

13.5.3　市场营销分析——市场页面广告统计模块介绍　/　282

13.5.4　市场营销分析——市场页面广告模块实现　/　283

13.6　恶意登录监控　/　286

13.6.1　恶意登录监控模块描述　/　286

13.6.2　恶意登录监控模块实现　/　286

13.7　订单支付实时监控　/　293

13.7.1　订单支付实时监控模块介绍　/　293

13.7.2　订单支付实时监控模块实现　/　293

13.8　订单支付实时对账　/　297

13.8.1　订单支付实时对账模块介绍　/　297

13.8.2　订单支付实时对账模块实现　/　297

第1章

大数据简介

本章内容介绍大数据的概述、特点、相关技术和应用场景等内容,以及学习大数据的规划和建议。

1.1 大数据的概述

大数据是指数据量巨大、类型多样、处理速度快、价值高、来源广泛、质量不确定、安全性要求高的数据。大数据技术主要用于处理这些数据,包括分布式计算、存储、处理和分析等方面。大数据技术可以帮助企业从数据中发现规律和价值,为企业决策提供有力支持。

学习大数据的基本概念,包括大数据的定义、特点、应用场景等。

学习大数据的具体实现技术,包括分布式计算、存储、处理和分析等方面。可以选择一种或多种大数据技术进行深入学习和实践,例如 Hadoop、HBase、Spark、Flink 等。

1.2 大数据的特点

1)数据量大:大数据技术主要用于处理海量数据,这些数据通常具有 TB 或 PB 级别的规模,需要使用分布式计算和存储技术进行处理。

2)数据类型多样:大数据技术可以处理各种类型的数据,包括结构化数据、半结构化数据和非结构化数据等。

3)数据处理速度快:大数据技术可以通过并行计算和分布式存储等方式,实现高速的数据处理和分析,可以在短时间内处理大量数据。

4)数据价值高:大数据技术可以通过对数据进行深度挖掘和分析,发现数据中隐藏的规律和价值,为企业决策提供有力支持。

5)数据来源广泛:大数据技术可以处理来自各种数据源的数据,包括传感器数据、社交媒体数据、日志数据等。

6)数据质量不确定:由于数据来源广泛,数据质量也不尽相同,大数据技术需要具备处理不确定数据的能力。

7) 数据安全性要求高：大数据技术处理的数据通常包含敏感信息，需要具备高度的数据安全性和隐私保护能力。

1.3 大数据的应用领域

大数据技术在各个领域都有着广泛的应用，大数据应用正在逐渐改变我们的生活和工作方式，它已经成为现代企业和社会不可或缺的一部分。通过使用大数据，企业可以更好地了解客户需求、优化业务流程、提高决策效率等。以下是大数据技术的一些应用场景，希望能够吸引读者的兴趣。

1) 金融行业：大数据技术在金融行业中有着广泛的应用。通过对大量的金融数据进行分析，可以帮助金融机构更好地了解市场趋势和客户需求，从而制定更加精准的投资策略和风险控制方案。同时，大数据技术还可以用于反欺诈、信用评估、交易监控等方面，提高金融机构的运营效率和风险管理能力。

2) 医疗健康：大数据技术在医疗健康领域中也有着广泛的应用。通过对大量的医疗数据进行分析，可以帮助医疗机构更好地了解疾病的发病机制和治疗效果，从而提高医疗服务的质量和效率。同时，大数据技术还可以用于疾病预测、个性化治疗、医疗资源调配等方面，为人们的健康保驾护航。

3) 零售行业：大数据技术在零售行业中也有着广泛的应用。通过对大量的销售数据进行分析，可以帮助零售企业更好地了解消费者的需求和购买行为，从而制定更加精准的营销策略和商品推荐方案。同时，大数据技术还可以用于库存管理、供应链优化、反欺诈等方面，提高零售企业的运营效率和风险管理能力。

4) 交通运输：大数据技术在交通运输领域中也有着广泛的应用。通过对大量的交通数据进行分析，可以帮助交通管理部门更好地了解交通状况和交通流量，从而制定更加精准的交通管理策略和交通规划方案。同时，大数据技术还可以用于交通安全预警、智能导航、公共交通优化等方面，提高交通运输的效率和安全性。

5) 教育行业：大数据技术在教育行业中也有着广泛的应用。通过对大量的教育数据进行分析，可以帮助教育机构更好地了解学生的学习情况和学习需求，从而制定更加精准的教学计划和教学评估方案。同时，大数据技术还可以用于学生评估、教育资源调配、在线教育等方面，提高教育服务的质量和效率。

以上是大数据技术的一些应用场景，它们各自具有不同的特点和作用。在实际应用中，根据项目需求和数据类型选择合适的技术进行数据处理和分析，可以提高数据处理和分析的效率和准确性。

1.4 目前企业应用的主流大数据技术

大数据技术正逐渐成为企业和组织的核心竞争力。大数据技术是用于存储、处理和分析大规模数据的技术，包括数据采集、数据存储、数据处理和数据分析等环节。通过掌握大数据技术，您将能够从海量数据中提取有价值的信息，以支持更加明智、精准的决策和业务操作。在本文中，我们将向您介绍一些基于大数据主流技术的核心概念和应用场景。常见的大数据技术包括以下几种。

1）Hadoop 是一个开源的分布式计算平台，用于存储和处理大规模数据。它的核心组件包括 HDFS 和 MapReduce。HDFS 是一个分布式文件系统，用于存储大规模数据。MapReduce 是一个分布式计算框架，用于处理大规模数据。Hadoop 的作用包括数据存储、数据处理和数据分析等方面。

2）Spark 是一个快速的、通用的分布式计算系统，用于大规模数据处理。它支持多种编程语言，包括 Java、Scala 和 Python 等。Spark 的核心组件包括 Spark Core、Spark SQL、Spark Streaming 和 MLlib 等。Spark 的作用包括数据处理、数据分析和机器学习等方面。

3）Flink 是一个快速的、可扩展的、分布式流处理框架，用于实时数据处理。它支持多种数据源，包括 Kafka、HDFS 和 Cassandra 等。Flink 的核心组件包括 DataStream API 和 Table API 等。Flink 的作用包括实时数据处理、实时数据分析和实时机器学习等方面。

4）Hive 是一个基于 Hadoop 的数据仓库工具，用于数据存储和查询。它支持 SQL 查询语言，可以将 SQL 语句转换为 MapReduce 任务进行执行。Hive 的作用包括数据存储和数据查询等方面。

5）HBase 是一个开源的文档型 NoSQL 数据库，用于存储大规模数据。它支持高可用性、高性能和可扩展性等特性。HBase 的作用包括数据存储和数据查询等方面。

以上是常见的大数据技术，它们各自具有不同的特点和作用。根据项目需求和数据类型选择合适的技术进行数据处理和分析，可以提高数据处理和分析的效率和准确性。

6）Kafka 是一种高吞吐量的分布式发布订阅消息系统，用于处理实时数据流。它支持多个生产者和消费者，并且可以自动进行数据分区和负载均衡。Kafka 还支持数据持久化和数据复制等功能。

在实际的大数据应用中，这些技术并不是孤立的，而是需要相互结合和集成。通过对这些技术的合理运用，您可以构建出高效、可靠的大数据解决方案，以支持企业的业务决策和发展。

希望通过本文的介绍，您对大数据主流技术有了初步的了解。接下来，您可以根据实际需求选择合适的技术和方法，将大数据技术应用到您的业务场景中，以实现更大的商业价值和发展潜力。

1.5 大数据开发流程

大数据开发流程是一套涉及多个阶段的综合性过程，旨在从原始数据中提取有价值的信息，支持业务决策和优化。下面我们将介绍大数据开发流程的主要阶段，帮助您更好地了解和掌握大数据应用的开发过程，我们可以将其分为以下几个步骤。

1）确定项目需求和目标：在大数据开发的流程中，首先需要明确项目的需求和目标。只有明确了项目的需求和目标，才能有针对性地进行数据收集和处理。这一步非常重要，因为它直接影响到后续的数据处理和分析。

2）收集和清洗数据：接下来，需要收集和清洗数据。数据的质量对后续的数据处理和分析至关重要，因此需要对数据进行清洗和预处理，以确保数据的准确性和完整性。

3）存储数据到大数据平台中：存储数据到大数据平台中是下一步。常见的大数据平台包括 Hadoop、Spark 和 Flink 等。这些平台可以用于存储、处理和分析大数据。在存储数据时，需要考虑数据的安全性和可扩展性等因素。

4）对数据进行处理和分析：对数据进行处理和分析是大数据开发的核心部分。在这一步中，需要使用编程语言和大数据平台等工具对数据进行处理和分析。常用的编程语言包括 Python、Java 等。常用的大数据平台包括 Hadoop、Spark、Flink 等。在数据处理和分析时，需要考虑数据的复杂性和处理速度等因素。

5）可视化和呈现数据：可视化和呈现数据是为了更好地展示数据分析结果。常用的数据可视化工具包括 Tableau、Power BI 等。通过数据可视化，可以更加直观地展示数据分析结果，帮助决策者更好地理解数据。

6）部署和维护项目：最后，需要部署和维护项目。在部署和维护项目时，需要考虑项目的可靠性和安全性等因素。常用的部署方式包括云计算和本地部署等。

作为一名大数据开发者，需要具备多种技能，包括数据处理和分析技能、编程技能、数据库技能、统计学和数学技能以及沟通和团队合作能力。如果你想成为一名大数据开发者，建议制定一个学习规划，包括学习大数据平台、编程语言、数据库、统计学和数学等方面的知识。希望这些信息能够帮助你更好地了解大数据开发！

第 2 章

大数据基础架构 Hadoop

本章内容介绍 Hadoop 的基本概念和工作原理，包括 HDFS 分布式文件系统、MapReduce 计算模型、YARN 资源管理器等。学习 Hadoop 的安装和配置，集群模式的部署。掌握 Hadoop 的常用命令和操作，包括文件上传、下载、删除，MapReduce 作业提交等操作。实践 Hadoop 的应用案例，包括大数据处理、统计分析等场景。

2.1 Hadoop 简介

Hadoop 是一个开源的分布式计算平台，用于存储和处理大规模数据集。Hadoop 的核心组件包括：分布式文件系统（Hadoop Distributed File System，简称 HDFS）、分布式资源管理（Yet Another Resource Negotiato，简称 YARN）和分布式计算框架（MapReduce）。Hadoop 的特点如下。

1）处理大规模数据集：Hadoop 可以处理大规模数据集，包括 PB 级别的数据。它可以自动管理集群中的资源使用情况，并根据作业的需求分配和调度任务，从而提高集群的利用率和性能。

2）分布式计算：Hadoop 是一个分布式计算框架，它可以将作业分配给不同的节点进行计算，从而加快计算速度。它还可以自动处理节点故障和数据备份，保证计算的可靠性和容错性。

3）易于扩展：Hadoop 可以很容易地扩展到数千个节点，从而处理更大规模的数据集。它还可以与其他开源工具和技术集成，例如 Hive、Pig、Spark 等。

4）开源免费：Hadoop 是一个开源的框架，可以免费使用和修改。它还有一个庞大的社区支持，可以提供技术支持和解决方案。

5）生态系统丰富：除了 Hadoop 本身，还有许多其他的开源工具和技术可以与之集成，例如 Hive、Spark 等。这些工具可以帮助我们更方便地处理数据，提取有价值的信息和知识。

2.2 Hadoop 架构详解

Hadoop 作为一个开源的分布式计算框架，具有高可靠性、高性能和灵活性等优势，被广泛应用于大数据存储、处理和分析的各个领域。Hadoop 架构包括多个组件和模块，可以灵活地组合和扩展，以满足不同场景的数据处理需求。

基于 Hadoop 架构的大数据解决方案可以帮助企业构建高效、可扩展的数据处理和分析平台，以应对日益增长的数据挑战。通过掌握 Hadoop 技术，您可以更好地应对大数据时代的挑战，为企业或组织提供更优质的数据支持和决策依据。

Hadoop 架构三大核心组件分别为：分布式存储（HDFS）、分布式资源管理（YARN）、离线计算（MapReduce），如图 2-1 所示。

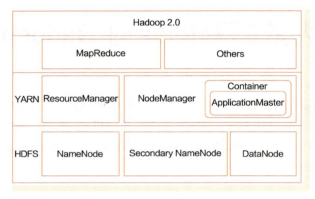

● 图 2-1 Hadoop 架构

2.2.1 分布式存储系统 HDFS

1. HDFS 简介

在大数据时代，数据量呈爆炸式增长，如何高效地存储和管理这些数据成了一个重要的问题。HDFS 作为 Hadoop 生态系统中的核心组件之一，为解决这个问题提供了一种可靠的、可扩展的分布式存储方案。

HDFS 是 Hadoop Distributed File System 的缩写，它是一个高度容错性的系统，被设计用来存储和分析大规模的数据集。HDFS 可以在商用服务器上存储和管理海量数据，并且具有高可靠性、高性能和可扩展性。

HDFS 主要由 NameNode、Secondary NameNode 和 DataNode 三个核心模块组成，它们之间的关系如图 2-2 所示。

2. HDFS 的核心组成

1）NameNode（简称 NN）：NameNode 是 HDFS 的主节点，它负责管理文件系统的命名空间和客户端对文件的访问。它的主要作用是维护文件系统的元数据，包括文件名、文件属性、文件目录结构等。

2）DataNode（简称 DN）：DataNode 是 HDFS 的数据节点，它负责存储文件数据和处理客户端的读写请求。它的主要作用是存储文件数据块，并向客户端提供数据读写服务。

3）Secondary NameNode（简称 SNN）：Secondary NameNode 是 NameNode 的辅助节点，它负责定期

合并和压缩 NameNode 的编辑日志，并生成新的镜像文件。它的主要作用是提高文件系统的可靠性和稳定性。

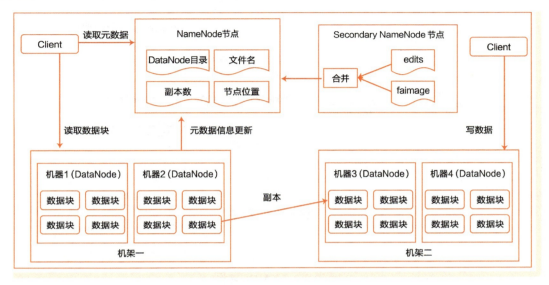

● 图 2-2　HDFS 架构

2.2.2　分布式资源管理框架 YARN

1. YARN 简介

在大数据应用不断扩展的今天，有效地管理和调度计算资源变得越来越重要。YARN 作为 Hadoop 生态系统中的核心组件之一，为解决这个问题提供了一种可扩展、高效的资源管理框架。

YARN（Yet Another Resource Negotiator）是一种新的 Hadoop 资源管理器，它是一个通用资源管理系统，负责管理和调度 Hadoop 集群中的资源。YARN 采用了分层的结构，主要由以下几个组件组成：ResourceManager、ApplicationMaster、NodeManager 和 Container 几个模块。它们之间的关系，如图 2-3 所示。

2. YARN 的核心组成以及它们的作用

1）ResourceManager：ResourceManager 是 YARN 的主节点，它负责管理集群中的资源并分配给不同的应用程序。它的主要作用是协调和监控集群中的资源使用情况，以及为应用程序提供资源分配和调度服务。ResourceManager 会根据应用程序的需求，将集群中的资源分配给 ApplicationMaster，然后由 ApplicationMaster 进一步分配给 Container。

2）ApplicationMaster：ApplicationMaster 是每个应用程序的主节点，它负责协调应用程序的执行。它的主要作用是向 ResourceManager 请求资源，并将资源分配给 Container，然后监控和管理 Container 的执行。ApplicationMaster 还负责处理应用程序的失败和重试，以及与 ResourceManager 和 NodeManager 之间的通信。

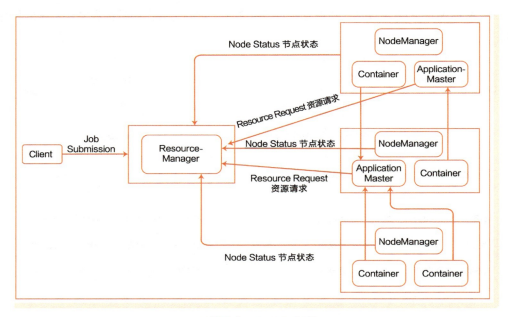

● 图 2-3　YARN 架构

3）NodeManager：NodeManager 是 YARN 的节点管理器，它负责管理单个节点上的资源并执行应用程序的任务。它的主要作用是监控节点上的资源使用情况，以及为应用程序提供资源分配和任务执行服务。NodeManager 会启动和停止 Container，并向 ApplicationMaster 报告 Container 的状态和资源使用情况。

4）Container：Container 是 YARN 的资源分配和执行单元，它是应用程序的一个实例。每个 Container 都有自己的资源和环境，包括内存、CPU、磁盘和网络等。Container 由 NodeManager 启动和停止，并由 ApplicationMaster 分配和管理。

要使用一个 YARN 集群，首先需要一个包含应用程序的客户的请求。ResourceManager 协商一个容器的必要资源，启动一个 ApplicationMaster 来表示已提交的应用程序。通过使用一个资源请求协议，ApplicationMaster 协商每个节点上供应用程序使用的资源容器。执行应用程序时，ApplicationMaster 监视容器直到完成。当应用程序完成时，ApplicationMaster 从 ResourceManager 注销其容器，执行周期就完成了。YARN 的原理详情请看本章的 2.5。

▶▶ 2.2.3　离线计算框架 MapReduce

1. MapReduce 简介

在大数据时代，数据量的迅速增长和计算任务的复杂性给传统的计算方法带来了巨大的挑战。MapReduce 作为 Hadoop 生态系统中的核心组件之一，为处理大规模数据提供了一种高效的离线计算框架。

MapReduce 是 Hadoop 的离线计算组件，是一种并行编程模型，用于大规模数据集的并行计算。

MapReduce 框架将并行计算抽象成为两个函数：Map 和 Reduce。MapReduce（离线计算框架）的组成如图 2-4 所示。

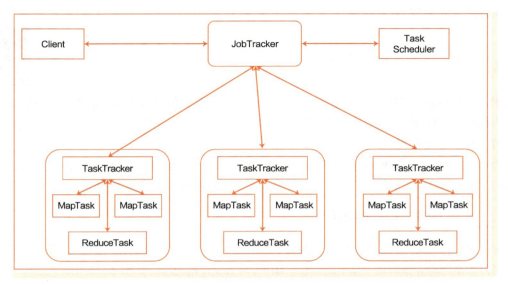

● 图 2-4　MapReduce 框架图

2. MapReduce 的核心组成及它们的作用

1）JobTracker：JobTracker 是 MapReduce 的主节点，它负责管理 MapReduce 作业并分配给不同的 TaskTracker。它的主要作用是协调和监控 MapReduce 作业的执行，以及为 TaskTracker 提供任务分配和调度服务。JobTracker 会根据作业的需求，将作业分配给 TaskTracker，然后由 TaskTracker 进一步分配给 Task。

2）TaskTracker：TaskTracker 是 MapReduce 的节点管理器，它负责管理单个节点上的任务并执行 MapReduce 作业的任务。它的主要作用是监控节点上的任务执行情况，以及为 JobTracker 提供任务分配和任务执行服务。TaskTracker 会启动和停止 Task，并向 JobTracker 报告 Task 的状态和资源使用情况。

3）Task：MapTask 和 ReduceTask 是 MapReduce 的任务执行单元，它们分别负责 Map 和 Reduce 阶段的计算。MapTask 将输入数据划分为若干个片段，并将每个片段分配给不同的 Mapper 进行计算。ReduceTask 将 Mapper 的输出结果按照 Key 进行分组，并将每个分组分配给不同的 Reducer 进行计算。MapTask 和 ReduceTask 都由 TaskTracker 启动和停止，并由 JobTracker 分配和管理。

4）InputFormat 和 OutputFormat：InputFormat 和 OutputFormat 是 MapReduce 的输入输出格式，它们负责将输入数据和输出数据转换为 MapReduce 作业所需的格式。InputFormat 将输入数据划分为若干个片段，并将每个片段分配给不同的 Mapper 进行计算。OutputFormat 将 Reducer 的输出结果写入到指定的输出目录中。

总之，MapReduce 通过 JobTracker、TaskTracker、MapTask、ReduceTask、InputFormat 和 OutputFormat

6 个组成部分，实现了高效的离线计算服务。它可以自动管理集群中的资源使用情况，并根据作业的需求分配和调度任务，从而提高集群的利用率和性能。同时，MapReduce 还提供了灵活的编程接口，可以支持各种类型的离线计算任务。

2.3 HDFS 读、写文件流程

HDFS 是 Hadoop 分布式存储系统的简称，它是一种高可靠性的分布式文件系统，可以存储和管理大规模的数据。在 HDFS 中，文件被分成多个数据块（block），每个数据块存储在多个节点上，这使得文件可以被分布式地存储和处理。

作为一个文件系统，文件的读和写是最基本的需求，客户端是如何与 HDFS 进行交互的，也就是客户端与 HDFS 以及构成 HDFS 的两类节点（NameNode 和 DataNode）之间的数据交互过程。

▶▶ 2.3.1 HDFS 写文件流程

学习 HDFS 写入文件的过程，您可以深入了解 HDFS 的基本原理和操作方法。这将为您在分布式环境下进行数据处理和分析打下坚实的基础。HDFS 写入文件的过程，如图 2-5 所示。

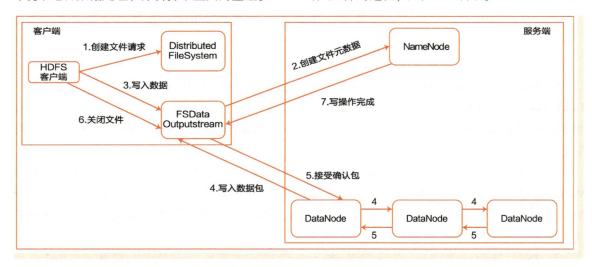

● 图 2-5 HDFS 写入文件过程

HDFS 写入文件的步骤

1）使用 HDFS Client，向远程的 NameNode 发起上传文件请求。

2）NameNode 会检查要创建的文件是否已经存在，若不存在则会为文件创建一个记录。

3）当客户端开始写入文件的时候，客户端会将文件切分成多个数据包，在内部以数据队列的形式管理这些数据包，并向 NameNode 申请的 DataNode 列表中写入数据。

4）开始以管道的形式将数据包写入第一个 DataNode，该 DataNode 把该数据包存储之后，再将其传递给在此管道中的下一个 DataNode，直到最后一个 DataNode，这种写数据的方式呈流水线的形式。

5）DataNode 成功存储之后会返回一个确认队列，在管道里传递至 HDFS Client，HDFS Client 成功收到 DataNode 返回的确认队列后，会移除相应的数据包。

6）传输过程完成后，会被关闭文件。

7）客户端完成数据的写入后，会对数据流调用 close()方法，关闭数据流。

2.3.2 HDFS 读文件流程

在 HDFS 中，文件被分成多个数据块（block），每个数据块存储在多个节点上，这使得文件可以被分布式地存储和处理。当我们需要从 HDFS 中读取文件时，可以使用 Hadoop 的 FileSystem API 或者 Hive 的 SELECT 语句等工具来完成。

通过学习 HDFS 读取文件的过程，您可以深入了解 HDFS 的基本原理和操作方法。这将帮助您更好地理解和应用 HDFS，从而更好地满足您的数据处理和分析需求。HDFS 读取文件的过程，如图 2-6 所示。

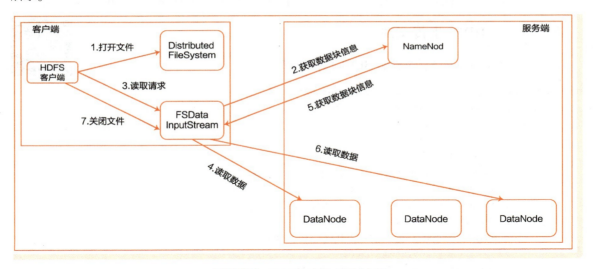

● 图 2-6　HDFS 读取文件过程

HDFS 读取文件的步骤如下。

1）HDFS Client 调用 FileSystem 实例的 open()方法，获得这个文件对应的输入流。

2）通过 RPC 远程调用 NameNode，获得 NameNode 中此文件对应的数据块保存位置，包括这个文件副本的保存位置（也就是各个 DataNode 的地址）。

3）获得输入流之后，HDFS Client 调用 read()方法读取数据。选择最近的 DataNode 建立连接并读取数据。

4）如果客户端和其中一个 DataNode 位于同一机器，那么就会直接从本地读取数据。

5）到达数据块末端，关闭与这个 DataNode 的连接，然后重新查找下一个数据块。

6）客户端调用 close()，关闭输入流。

2.4 HDFS 的实战操作

在 Hadoop 生态系统中，HDFS 扮演着至关重要的角色，它为大数据分析和处理提供了基础架构支持。在实战操作中，我们将通过使用 HDFS 的 API 来演示如何进行文件的读写操作。我们将使用 Java 语言来实现这些操作，因为 Java 是 Hadoop 生态系统中最常用的编程语言之一。

为了编写 HDFS 的 API 实践操作文件，你需要进行以下步骤。

1) 必须要先启动 Hadoop 集群，配置好环境等。

2) 在 Windows 系统需要设置 Hadoop 环境，本节内容使用已经编译好的 Windows 版本 hadoop-common-2.2.0-bin-master，需要做以下配置。

3) 在 Windows 系统的高级系统设置配置选项中配置 HADOOP_HOME 指向 hadoop-common-2.2.0-bin-master 安装包目录。

4) 在 Windows 系统的高级系统设置配置选项中 path 变量中加入 HADOOP_HOME 的 bin 目录。

2.4.1 HDFS 上传文件

以下是使用 Java 上传本地文件到 HDFS 的示例代码。

```java
import org.apache.hadoop.conf.Configuration;
import org.apache.hadoop.fs.FileSystem;
import org.apache.hadoop.fs.Path;

public class HdfsUploader {
    public static void main(String[] args) throws Exception {
        // Create a configuration object
        Configuration conf = new Configuration();
        // Set the URI of the HDFS cluster
        conf.set("fs.defaultFS", "hdfs://localhost:9000");
        // Create a FileSystem object
        FileSystem fs = FileSystem.get(conf);
        // Specify the path of the local file
        Path localFilePath = new Path("/path/to/local/file");
        // Specify the path of the HDFS file
        Path hdfsFilePath = new Path("/path/to/hdfs/file");
        // Upload the local file to HDFS
        fs.copyFromLocalFile(localFilePath, hdfsFilePath);
        // Close the FileSystem object
        fs.close();
    }
}
```

在这段代码中，我们首先创建了一个 Configuration 对象，并设置了 HDFS 集群的 URI。然后，我们使用 FileSystem 类的 get 方法创建了一个 FileSystem 对象。我们使用 Path 类指定了本地文件和 HDFS 文件的路径。最后，我们使用 FileSystem 对象的 copyFromLocalFile 方法将本地文件上传到 HDFS。

▶▶ 2.4.2 HDFS 创建文件

以下是使用 Java 创建 HDFS 文件的示例代码。

```java
import org.apache.hadoop.conf.Configuration;
import org.apache.hadoop.fs.FileSystem;
import org.apache.hadoop.fs.Path;

public class HdfsFileCreator {
    public static void main(String[] args) throws Exception {
        // 创建一个配置对象
        Configuration conf = new Configuration();
        // 设置 HDFS 集群的 URI
        conf.set("fs.defaultFS", "hdfs://localhost:9000");
        // 创建一个 FileSystem 对象
        FileSystem fs = FileSystem.get(conf);
        // 指定 HDFS 文件的路径
        Path hdfsFilePath = new Path("/path/to/hdfs/file");
        // 创建一个空的 HDFS 文件
        fs.create(hdfsFilePath);
        // 关闭 FileSystem 对象
        fs.close();
    }
}
```

在这段代码中，我们首先创建了一个 Configuration 对象，并设置了 HDFS 集群的 URI。然后使用 FileSystem 类的 get 方法创建了一个 FileSystem 对象。使用 Path 类指定了 HDFS 文件的路径。最后，使用 FileSystem 对象的 create 方法创建了一个空的 HDFS 文件。

▶▶ 2.4.3 HDFS 创建目录

以下是使用 Java 创建 HDFS 目录的示例代码。

```java
import org.apache.hadoop.conf.Configuration;
import org.apache.hadoop.fs.FileSystem;
import org.apache.hadoop.fs.Path;

public class HdfsDirectoryCreator {
    public static void main(String[] args) throws Exception {
        // 创建一个配置对象
        Configuration conf = new Configuration();
        // 设置 HDFS 集群的 URI
        conf.set("fs.defaultFS", "hdfs://localhost:9000");
        // 创建一个 FileSystem 对象
        FileSystem fs = FileSystem.get(conf);
```

```
        // 指定 HDFS 目录的路径
        Path hdfsDirectoryPath = new Path("/path/to/hdfs/directory");
        // 创建一个 HDFS 目录
        fs.mkdirs(hdfsDirectoryPath);
        // 关闭 FileSystem 对象
        fs.close();
    }
}
```

在这段代码中,我们首先创建了一个 Configuration 对象,并设置了 HDFS 集群的 URI。然后,使用 FileSystem 类的 get 方法创建了一个 FileSystem 对象。使用 Path 类指定了 HDFS 目录的路径。最后,使用 FileSystem 对象的 mkdirs 方法创建了一个 HDFS 目录。

2.4.4 HDFS 重命名文件

以下是使用 Java 重命名 HDFS 文件的示例代码。

```
import org.apache.hadoop.conf.Configuration;
import org.apache.hadoop.fs.FileSystem;
import org.apache.hadoop.fs.Path;

public class HdfsFileRenamer {
    public static void main(String[] args) throws Exception {
        // 创建一个配置对象
        Configuration conf = new Configuration();
        // 设置 HDFS 集群的 URI
        conf.set("fs.defaultFS", "hdfs://localhost:9000");
        // 创建一个 FileSystem 对象
        FileSystem fs = FileSystem.get(conf);
        // 指定原始 HDFS 文件的路径
        Path oldFilePath = new Path("/path/to/old/hdfs/file");
        // 指定新的 HDFS 文件的路径
        Path newFilePath = new Path("/path/to/new/hdfs/file");
        // 重命名 HDFS 文件
        fs.rename(oldFilePath, newFilePath);
        // 关闭 FileSystem 对象
        fs.close();
    }
}
```

在这段代码中,我们首先创建了一个 Configuration 对象,并设置了 HDFS 集群的 URI。然后,使用 FileSystem 类的 get 方法创建了一个 FileSystem 对象。使用 Path 类指定了原始 HDFS 文件和新的 HDFS 文件的路径。最后,使用 FileSystem 对象的 rename 方法重命名了 HDFS 文件。

2.4.5 HDFS 删除文件

以下是使用 Java 删除 HDFS 文件的示例代码。

```java
import org.apache.hadoop.conf.Configuration;
import org.apache.hadoop.fs.FileSystem;
import org.apache.hadoop.fs.Path;

public class HdfsFileDeleter {
    public static void main(String[] args) throws Exception {
        // 创建一个配置对象
        Configuration conf = new Configuration();
        // 设置 HDFS 集群的 URI
        conf.set("fs.defaultFS", "hdfs://localhost:9000");
        // 创建一个 FileSystem 对象
        FileSystem fs = FileSystem.get(conf);
        // 指定要删除的 HDFS 文件的路径
        Path hdfsFilePath = new Path("/path/to/hdfs/file");
        // 删除 HDFS 文件
        fs.delete(hdfsFilePath, true);
        // 关闭 FileSystem 对象
        fs.close();
    }
}
```

在这段代码中,我们首先创建了一个 Configuration 对象,并设置了 HDFS 集群的 URI。然后,使用 FileSystem 类的 get 方法创建了一个 FileSystem 对象。使用 Path 类指定了要删除的 HDFS 文件的路径。最后,使用 FileSystem 对象的 delete 方法删除了 HDFS。

2.5 YARN 原理

1. YARN 原理介绍

YARN 是一种先进的资源管理系统,用于在集群中对多个应用程序进行资源管理和调度。YARN 的出现极大地扩展了 Hadoop 的应用范围和能力,使得它不再仅仅是一个只适用于处理大规模数据集的分布式文件系统。它可以将集群中的资源(如 CPU 和内存)分配给各个任务,从而实现高效的资源管理和调度。YARN 工作机制,如图 2-7 所示。

在 YARN 中,应用程序的提交包括打包和配置应用程序代码,并使用 YARN 客户端将应用程序提交到 YARN 资源管理器。在提交过程中,我们需要指定应用程序的名称、运行环境、资源需求等信息。

YARN 资源管理器将根据应用程序的资源需求为其分配相应的资源。资源分配的过程包括选择合适的节点或节点群,并在这些节点上启动应用程序的容器。资源管理器将根据节点的资源使用情况和应用程序的需求进行动态调度,以确保资源的合理分配和应用程序的性能。

2. YARN 工作的步骤

1) MapReduce 程序提交到客户端所在节点,然后 Yarn Runner 向 ResourceManager 申请一个 Application 资源。

2) ResourceManager 返回给客户端提交路径以及 ApplicationId。

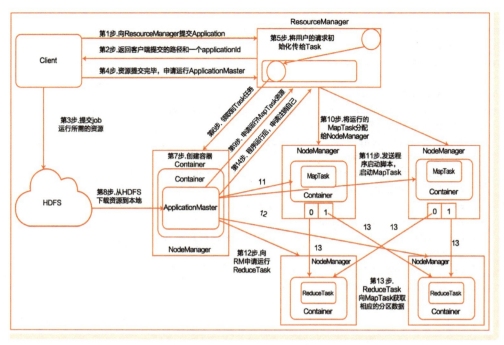

● 图 2-7 YARN 工作机制

3）YarnRunner 提交 job 运行所需资源，包括该 job 所需切片的信息、job 在 Hadoop 集群中的参数配置信息和使用的 jar 包。

4）资源提交完成后，YarnRunner 向 ResourceManager 申请运行 ApplicationMaster。

5）ResourceManager 会在内部将用户的请求初始化一个 Task，然后放入任务队列里面等待执行。

6）等到 NodeManager 空闲后领取到 Task 任务便创建 Container 容器。

7）Container 容器在里面启动 ApplicationMaster。

8）Container 容器读取 job 资源，获取到了 job 切片信息。

9）向 ResourceManager 申请 MapTask 容器用来执行 MapTask 任务。其他空闲 NodeManager 空闲后领取任务，创建对应切片个数的 Container 容器。

10）AppMaster 发送程序脚本启动对应的 MapReduce 任务，即为 Map 任务进程。

11）当 Map 任务运行完成落盘之后，MapReduce 发送启动程序脚本，启动 MapTask。

12）MapReduce 会再次向 ResourceManager 申请执行 ReduceTask 任务的资源。

13）当 ReduceTask 任务也运行完成之后，MapReduce 通知 ResourceManager 并注销自己，同时相关的 MapReduce 的资源也释放掉。

2.6 YARN 调度器详解

1. YARN 调度介绍

调度器是 YARN 中的关键组件，它负责在集群中分配资源给正在运行的应用程序。调度器根据应用程序的需求和集群的资源使用情况来动态调度资源,以确保资源的合理分配和应用程序的性能。在 YARN 中,调度器采用了多种策略来进行资源调度,如 FIFO(先进先出)、公平调度和容量调度等。这些策略各有优劣,适用于不同的场景和需求。

2. YARN 调度器分类

YARN 调度器主要分为三类:FIFO Scheduler(先进先出调度器)、Capacity Scheduler(容量调度器)和 Fair Sceduler(公平调度器)。

(1)FIFO Scheduler 分析

FIFO Scheduler 就像数据结构中的 FIFO(First in first out)一样,在调度器里 FIFO Scheduler 也是一个先进先出的队列,如图 2-8 所示。

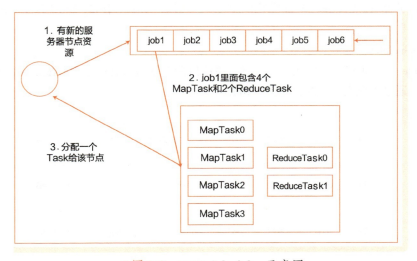

• 图 2-8 FIFO Scheduler 示意图

FIFO Scheduler 把任务按照提交的顺序排成一个队列,在进行资源分配的时候,先把资源分配给最顶部的任务,只有最顶部的任务的资源需求得到了满足,才会分配给下一个任务。因此,会导致其他任务被阻塞,通常情况下同一时间只有一个任务在执行,在实际生产环境里一般不会用这种调度器。

(2)Capacity Scheduler 分析

默认的资源调度器。会把任务分为多个队列,同时每个队列内部先进先出,同一时间队列中只有一个任务在执行,队列的并行度等于队列的个数,如果使用容量调度器,就可以进行 YARN 资源队列配置,如图 2-9 所示。

Capacity Scheduler 的特点。

1)支持多个队列,每个队列可配置一定的资源量,每个队列采用 FIFO 调度策略。

2)为了防止同一个用户的作业独占队列中的资源,该调度器会对同一用户提交的作业所占资源量进行限定。

3)首先,计算每个队列中正在运行的任务数与其应该分得的计算资源之间的比值,选择一个该比值最小的队列(最闲的)。

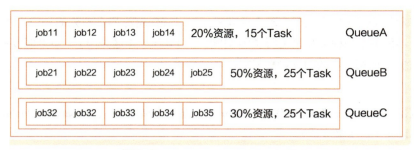

● 图 2-9　Capacity Scheduler 示意图

4）其次，按照作业优先级和提交时间顺序，同时考虑用户资源量限制和内存限制对队列内任务排序。

5）三个队列同时按照任务的先后顺序依次执行，比如，job11、job21 和 job31 分别排在队列最前面，先运行，也是并行运行。

（3）Fair Scheduler 分析

Fair Scheduler 公平调度器和它的名字一样，设计目标就是为所有的应用分配公平的资源，而这个公平是我们可以通过参数来设置的，如图 2-10 所示。

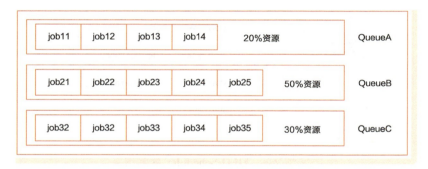

● 图 2-10　Fair Scheduler 示意图

它也是有多个队列，但是每个队列内部是按照缺额大小来分配资源并启动任务的。也就是同一队列中的作业公平共享队列中所有资源，同一时间队列中可以有多个任务执行，所以队列的并行度大于等于队列的个数。

2.7　MapReduce 工作原理

1. MapReduce 原理介绍

MapReduce 是一种编程模型，用于处理和生成大型数据集。通过 MapReduce，开发人员可以编写简单的程序，以并行处理大规模数据集，并自动处理集群中的任务调度、容错和负载均衡等问题。

MapReduce 工作原理，主要分为 MapTask 和 ReduceTask，以单词统计为例，展示 MapReduce 内部工作流程和细节，如图 2-11 所示。

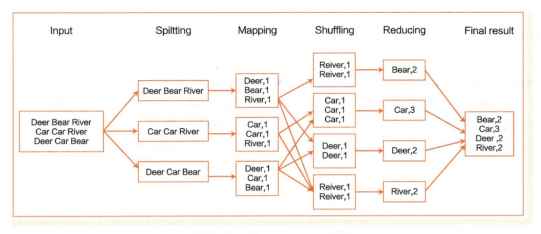

● 图 2-11　MapReduce 工作原理

2. MapReduce 的运行流程

MapReduce 的运行流程大致分为：输入→map→shuffle→reduce→输出，下面介绍 MapReduce 的过程。

1）MapReduce 程序获取输入和输出的目录。

2）客户端将应用程序提交到 Hadoop 集群中，使用 InputFormat 组件将数据切分成若干分片。

3）MapTask 将对应的分片，每一行解析成一个 key-value 对，并且调用 map( ) 函数处理。

4）处理完成之后，根据 ReduceTask 个数调用 partition 组件进行分区，然后进行排序和本地归约，写入本地的磁盘。

5）ReduceTask 会从网络上读取 MapTask 处理后的属于自己的分区（也就是 shuffle），把 key 相同的数据放在一起。

6）最后调用 reduce( ) 函数处理数据，并将处理的结果写到 HDFS 上。

3. MapTask 的运行流程

MapTask 流程是将数据打散，然后计算并将结果保存，具体步骤如下。

1）Read 阶段：MapTask 通过用户编写的 RecordReader，从输入 InputSplit 中解析出一个个 key-value。

2）Map 阶段：该阶段主要是将解析出的 key-value 交给用户编写的 map( ) 函数处理，并产生一系列新的 key-value。

3）Collect 阶段：在用户编写的 map( ) 函数中，当数据处理完成后，一般会调用 OutputCollector.collect( ) 输出结果。在该函数内部，它会将生成的 key-value 分片（通过调用 Partition），并写入一个环形内存缓冲区中。

4）Spill 阶段：即"溢写"，当环形缓冲区满后，MapReduce 会将数据写到本地磁盘上，生成一个临时文件。需要注意的是，将数据写入本地磁盘之前，先要对数据进行一次本地排序，并在必要时对数据进行合并操作。

5）Combine 阶段：当所有数据处理完成后，MapTask 对所有临时文件进行一次合并，以确保最终只会生成一个数据文件。

4. ReduceTask 的运行流程

ReduceTask 流程是将数据汇总，然后保存，具体步骤如下。

1）Shuffle 阶段：ReduceTask 从各个 MapTask 所在的 TaskTracker 上远程拷贝一片数据，并针对某一片数据，如果其大小超过一定阈值，则写到磁盘上，否则直接放到内存中。

2）Merge 阶段：在远程拷贝数据的同时，ReduceTask 启动了两个后台线程对内存和磁盘上的文件进行合并，以防止内存使用过多或磁盘上的文件过多，并且可以为后面整体的归并排序减负，提升排序效率。

3）Sort 阶段：按照 MapReduce 的语义，用户编写的 reduce( ) 函数输入数据是按 key 进行聚集的一组数据。为了将 key 相同的数据聚集在一起，Hadoop 采用了基于排序的策略。由于各个 MapTask 已经实现自己的处理结果并进行了局部排序，因此，ReduceTask 只需要对数据进行一次归并排序即可。

4）Reduce 阶段：在该阶段中，ReduceTask 将每组数据依次交给用户编写的 reduce( ) 函数处理。

5）Write 阶段：reduce( ) 函数将计算结果写到 HDFS 上。

2.8 MapReduce 核心的原理 Shuffle

Shuffle 本意是洗牌、混洗的意思，把一组有规则的数据尽量打乱成无规则的数据。而在 MapReduce 中，Shuffle 更像是洗牌的逆过程，指的是将 Map 端的无规则输出按指定的规则"打乱"成具有一定规则的数据，以便 Reduce 端接收处理。其在 MapReduce 中所处的工作阶段是 Map 输出后到 Reduce 接收前，具体可以分为 Map 端和 Reduce 端前后两个部分，如图 2-12 所示。

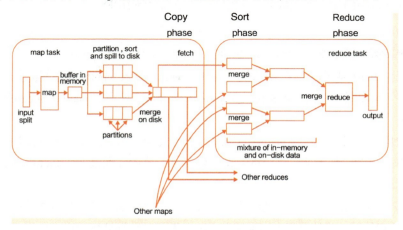

• 图 2-12　Shuffle 原理图

2.8.1 Map 端

1. Map 的 Shuffle 介绍

在 Map 端的 Shuffle 过程是对 Map 的结果进行分区、排序、分割，然后将属于同一分区的输出合并在一起并写在磁盘上，最终得到一个分区有序的文件，分区有序的含义是 Map 输出的键值对按分区进行排列，具有相同 partition 值的键值对存储在一起，每个分区里面的键值对又按 key 值进行升序排列，其流程如图 2-13 所示。

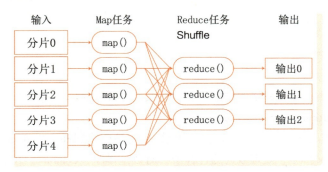

● 图 2-13　Map 端 Shuffle 过程示意图

2. Map 的 Shuffle 流程

1）在 Map 端首先是 InputSplit，在 InputSplit 中含有 DataNode 中的数据，每一个 InputSplit 都会分配一个 Mapper 任务，Mapper 任务结束后产生<Key，Value>的输出，这些输出先存放在缓存中，每个 Map 有一个环形内存缓冲区，用于存储任务的输出。默认大小 100MB，一旦达到阈值 0.8（即 80MB），一个后台线程就把内容写到（spill）本地磁盘中的指定目录下新建的一个溢出写文件。

2）写磁盘前，要进行 partition、sort 和 combine 等操作。通过分区，将不同类型的数据分开处理，之后对不同分区的数据进行排序，如果有 Combiner，还要对排序后的数据进行 combine。等最后记录写完，将全部溢出文件合并为一个分区且排序的文件。

3）最后将磁盘中的数据送到 Reduce 中，从图中可以看出 Map 输出有三个分区，有一个分区数据被送到图示的 Reduce 任务中，剩下的两个分区被送到其他 Reducer 任务中。而图示的 Reducer 任务的其他三个输入则来自其他节点的 Map 输出。

2.8.2 Reduce 端

Reduce 端的 Shuffle 流程

1）Copy 阶段：Reducer 通过 http 方式得到 Map 输出文件的分区。

Reduce 端可能从 n 个 Map 的结果中获取数据，而这些 Map 的执行速度不尽相同，当其中一个 Map 运行结束时，Reduce 就会从 JobTracker 中获取该信息。Map 运行结束后 TaskTracker 会得到消息，进而将消息汇报给 JobTracker，Reduce 定时从 JobTracker 获取该信息，Reduce 端默认有 5 个数据复制线程从 Map 端复制数据。

2) Merge 阶段：如果形成多个磁盘文件，会进行合并。

从 Map 端复制来的数据首先写到 Reduce 端的缓存中，同样缓存占用到达一定阈值后，会将数据写到磁盘中，同样会进行分区、合并、排序等过程。如果形成了多个磁盘文件，还会进行合并，最后一次合并的结果作为 Reduce 的输入，而不是写入到磁盘中。

3) Reducer 的参数：最后将合并后的结果作为输入传入 Reduce 任务中。

最后就是 Reduce 过程了，Reduce 过程将结果汇总，在这个过程中产生了最终的输出结果，并将其写到 HDFS 上。

2.9 MapReduce 常用三大组件

为了提高计算的效率，减少网络的传输，MapReduce 在 MapTask 阶段做了一些优化，内部嵌入了三大组件，分别是：Partitioner、Sort、Combiner。

Partitioner 组件的作用就是根据自定义方式确定数据将被传输到哪一个 Reducer 进行处理；Sort 组件的作用可以实现自定义排序，处理数据的过程中会对数据排序；Combiner 组件的作用在 MapTask 之后给 MapTask 的结果进行局部汇总，以减轻 reducetask 的计算负载，减少网络传输。

▶▶ 2.9.1 MapReduce 中的 Partitioner

Partitioner 组件在 MapReduce 的 MapTask 阶段对 Mapper 产生的中间结果进行分片，以便将同一分组的数据交给同一个 Reduce 处理，Partitioner 直接影响 Reduce 阶段的负载均衡。Partitioner 值默认是通过计算 key 的 hash 值后，对 Reduce Task 的数量取模获得。Mapreduce 通过 Partitioner 对 Key 进行分区，然后把数据按自己的需求来分发。

用户可以按自己的需求进行分组，如果想自定义数据分组规则，可以重写组件 Partitioner，自定义一个类继承抽象类 Partitioner，然后在 Job 对象中，设置自定义 Partitioner。

MapReduce 自定义分组类 ProvincePartitioner 继承 Partitione，实例代码如下：

```java
package com.jareny.bigdata.mapreduce.flow;

import java.util.HashMap;
import org.apache.hadoop.io.Text;
import org.apache.hadoop.mapreduce.Partitioner;
public class ProvincePartitioner extends
        Partitioner<Text, FlowBean> {
    private static HashMap<String, Integer>
        provincMap = new HashMap<String, Integer>();

    static {
        provincMap.put("138", 0);
        provincMap.put("139", 1);
```

```java
        provincMap.put("136", 2);
        provincMap.put("137", 3);
        provincMap.put("135", 4);
    }

    @Override
    public int getPartition(Text key, FlowBean value, int numPartitions) {
        Integer code = provincMap.get(key.toString().substring(0, 3));
        if (code != null) {
            return code;
        }
        return 5;
    }
}
```

MapReduce 自定义分组的实现逻辑,实例代码如下。

```java
package com.jareny.bigdata.mapreduce.sort;

import java.io.IOException;
import org.apache.commons.lang.StringUtils;
import org.apache.hadoop.conf.Configuration;
import org.apache.hadoop.fs.FileSystem;
import org.apache.hadoop.fs.Path;
import org.apache.hadoop.io.LongWritable;
import org.apache.hadoop.io.Text;
import org.apache.hadoop.mapreduce.Job;
import org.apache.hadoop.mapreduce.Mapper;
import org.apache.hadoop.mapreduce.Reducer;
import org.apache.hadoop.mapreduce.lib.input.FileInputFormat;
import org.apache.hadoop.mapreduce.lib.output.FileOutputFormat;

public class FlowSumProvince {

    // 重写 Mapper 类
    public static class FlowSumProvinceMapper
            extends Mapper<LongWritable, Text, Text, FlowBean> {
        Text k = new Text();
        FlowBean v = new FlowBean();

        @Override
        protected void map(LongWritable key, Text value,
            Context context) throws IOException, InterruptedException {

            // 将读到的一行数据进行字段切分
            String line = value.toString();
            String[] fields = StringUtils.split(line, "\t");
```

```java
        // 抽取业务所需要的各字段
        String phone = fields[1];
        long upFlow = Long.parseLong(fields[ fields.length -3]);
        long downFlow = Long.parseLong(fields[ fields.length -2]);
        k.set(phone);
        v.setFlowBean(upFlow, downFlow);
        context.write(k, v);
    }
}

// 重写 Reducer 类
public static class FlowSumProvinceReducer
        extends Reducer<Text, FlowBean, Text, FlowBean> {
    @Override
    protected void reduce(Text key, Iterable<FlowBean> values,
            Context context) throws IOException, InterruptedException {
        int upCount = 0;
        int downCount = 0;
        for (FlowBean bean : values) {
            upCount += bean.getUpFlow();
            downCount += bean.getDownFlow();
        }
        FlowBean sumBean = new FlowBean();
        sumBean.setFlowBean(upCount, downCount);
        context.write(key, sumBean);
    }
}

public static void main(String[] args) throws Exception {
    Configuration conf = new Configuration();
    Job job = Job.getInstance(conf);
    job.setJarByClass(FlowSumProvince.class);

    // 告诉框架,我们的程序所用的 mapper 类和 reducer 类
    job.setMapperClass(FlowSumProvinceMapper.class);
    job.setReducerClass(FlowSumProvinceReducer.class);
    job.setMapOutputKeyClass(Text.class);
    job.setMapOutputValueClass(FlowBean.class);

    // 告诉框架,我们的 mapper reducer 输出的数据类型
    job.setOutputKeyClass(Text.class);
    job.setOutputValueClass(FlowBean.class);

    // 设置 shuffle 的分区组件使用我们自定义的分区组件
    job.setPartitionerClass(ProvincePartitioner.class);
    // 设置 reduce task 的数量
```

```
        job.setNumReduceTasks(6);

        //告诉框架,我们要处理的文件在哪个路径下
        FileInputFormat.setInputPaths(job, new Path(args[0]));
        //告诉框架,我们的处理结果要输出到哪里去
        Path out = new Path(args[1]);
        FileSystem fs = FileSystem.get(conf);
        if(fs.exists(out)){
            fs.delete(out, true);
        }
        FileOutputFormat.setOutputPath(job,out);
        boolean res = job.waitForCompletion(true);
        System.exit(res ? 0 : 1);
    }
}
```

2.9.2 MapReduce 中的 Sort

MapReduce 程序在处理数据的过程中会对数据排序,排序的依据是 Map 输出的 key,也就是 MapReduce 本身自带排序,在 Map 阶段是局部有序,在 Reduce 阶段是全局有序的。Sort 组件的作用是提升 Map 阶段传输到 Reduce 阶段的数据统计处理。

用户可以根据业务需求,用自定义的 Bean 来封装数据排序,实现自己需要的排序规则。自定义的 Bean 来封装数据排序可以考虑将排序因素放到 key 中,让 key 实现接口 WritableComparable,然后重写 key 的 compareTo 方法。

MapReduce 的自定义排序 FlowBean 实现 WritableComparable 接口,实例代码如下。

```
package com.jareny.bigdata.mapreduce.flow;

import lombok.Data;
import org.apache.hadoop.io.WritableComparable;
import java.io.DataInput;
import java.io.DataOutput;
import java.io.IOException;
//  MapReduce 自定义排序的实现类
@Data
public class FlowBean implements WritableComparable<FlowBean> {
    private long upFlow;           //上行流量
    private long downFlow;         //下行流量
    private long totalFlow;        //总流量

    //按照总流量倒序排
    @Override
    public int compareTo(FlowBean bean) {
        return bean.totalFlow>this.totalFlow? 1:-1;
    }
```

```java
//序列化时需要无参构造方法
public FlowBean() {
}

public FlowBean(long upFlow, long downFlow) {
    this.upFlow = upFlow;
    this.downFlow = downFlow;
    this.totalFlow = upFlow + downFlow;
}

public void setFlowBean(long upFlow, long downFlow) {
    this.upFlow = upFlow;
    this.downFlow = downFlow;
    this.totalFlow = upFlow + downFlow;
}

//序列化方法 hadoop 的序列化很简单,要传递的数据写出去即可
@Override
public void write(DataOutput out) throws IOException {
    out.writeLong(upFlow);
    out.writeLong(downFlow);
    out.writeLong(totalFlow);
}

//反序列化方法注意:反序列化的顺序跟序列化的顺序完全一致
@Override
public void readFields(DataInput in) throws IOException {
    this.upFlow = in.readLong();
    this.downFlow = in.readLong();
    this.totalFlow = in.readLong();
}

//重写 toString 以便展示
@Override
public String toString() {
    return upFlow + "\t" + downFlow + "\t" + totalFlow;
}
}
```

使用 MapReduce 的自定义排序的 FlowBean 实现排序的逻辑,实例代码如下。

```java
package com.jareny.bigdata.mapreduce.flow;

import org.apache.hadoop.conf.Configuration;
import org.apache.hadoop.io.LongWritable;
import org.apache.hadoop.io.Text;
```

```java
import org.apache.hadoop.mapreduce.Job;
import org.apache.hadoop.mapreduce.Mapper;
import org.apache.hadoop.mapreduce.Reducer;
import org.apache.hadoop.fs.Path;
import org.apache.hadoop.mapreduce.lib.input.FileInputFormat;
import org.apache.hadoop.mapreduce.lib.output.FileOutputFormat;

import java.io.IOException;

public class FlowSumSort {
    public static class FlowSumSortMapper extends
            Mapper<LongWritable,Text,FlowBean,Text> {

        FlowBean k = new FlowBean();
        Text v = new Text();

        @Override
        protected void map(LongWritable key, Text value, Context context)
                throws IOException, InterruptedException {

            String line = value.toString();
            String[] fields = line.split("\t");
            String phone = fields[0];
            long upFlowSum = Long.parseLong(fields[2]);
            long dFlowSum = Long.parseLong(fields[3]);
            k.setFlowBean(upFlowSum, dFlowSum);
            v.set(phone);
            context.write(k, v);
        }
    }

    public static class FlowSumSortReducer extends
            Reducer<FlowBean, Text, Text, FlowBean> {
        @Override
        protected void reduce(FlowBean bean, Iterable<Text> phones, Context context)
                throws IOException, InterruptedException {
            context.write(phones.iterator().next(), bean);
        }
    }

    public static void main(String[] args)
            throws IOException, InterruptedException, ClassNotFoundException {
        Configuration conf = new Configuration();
        Job job = Job.getInstance(conf);
        job.setJarByClass(FlowSumSort.class);
```

```java
        // 告诉框架,我们的程序所用的 mapper 类和 reducer 类
        job.setMapperClass(FlowSumSortMapper.class);
        job.setReducerClass(FlowSumSortReducer.class);
        job.setMapOutputKeyClass(FlowBean.class);
        job.setMapOutputValueClass(Text.class);

        // 告诉框架,我们的 mapperreducer 输出的数据类型
        job.setOutputKeyClass(Text.class);
        job.setOutputValueClass(FlowBean.class);

        // 告诉框架,我们要处理的文件在哪个路径下
        FileInputFormat.setInputPaths(job, new Path(args[0]));
        // 告诉框架,我们的处理结果要输出到哪里去
        FileOutputFormat.setOutputPath(job, new Path(args[1]));

        boolean res = job.waitForCompletion(true);
        System.exit(res ? 0 : 1);
    }
}
```

2.9.3 MapReduce 中的 Combiner

Combiner 组件的作用在 MapTask 之后对 MapTask 的结果进行局部汇总,可以理解为 Map 任务的本地 Reduce 操作,以减轻 ReduceTask 的计算负载,减少网络传输。

用户可以根据业务的规则,优化计算减少网络传输,可以编写一个类,然后继承 Reducer,在 reduce( )方法中写具体的 Combiner 逻辑,然后在 Job 中设置 Combiner 组件。

MapReduce 自定义局部排序 FlowSumCombine 类继承 Reducer,实例代码如下。

```java
package com.jareny.bigdata.mapreduce.flow;

import org.apache.hadoop.io.Text;
import org.apache.hadoop.mapreduce.Reducer;
import java.io.IOException;

public class FlowSumCombine   extends
        Reducer<Text, FlowBean, Text, FlowBean> {

    FlowBean v = new FlowBean();
    // combiner 的逻辑和 reducer 的逻辑一样
    @Override
    protected void reduce(Text key, Iterable<FlowBean> values,Context context)
            throws InterruptedException, IOException {

        long upFlowCount = 0;
        long downFlowCount = 0;
```

```
        for (FlowBean bean : values) {
            upFlowCount += bean.getUpFlow();
            downFlowCount += bean.getDownFlow();
        }
        v.setFlowBean(upFlowCount, downFlowCount);
        context.write(key, v);
    }
}
```

2.10 MapReduce 项目实战

日志数据成了企业获取洞察和优化运营的重要来源。然而，这些日志通常包含大量噪声和冗余信息，需要进行清洗和过滤，以提取有价值的数据。本次实战将利用 MapReduce 进行日志清洗，通过分布式计算的方式对海量数据进行高效处理。

日志清洗项目旨在从 Web 服务器日志中提取有用的信息，如访问时间、IP 地址、请求 URL 等，为后续的数据分析和挖掘提供高质量的数据源。

2.10.1 清洗日志

使用 MapReduce 清洗日志，代码如下。

```
import java.io.IOException;
import java.util.Iterator;
import org.apache.hadoop.conf.Configuration;
import org.apache.hadoop.fs.Path;
import org.apache.hadoop.io.LongWritable;
import org.apache.hadoop.io.Text;
import org.apache.hadoop.mapreduce.Job;
import org.apache.hadoop.mapreduce.Mapper;
import org.apache.hadoop.mapreduce.Reducer;
import org.apache.hadoop.mapreduce.lib.input.FileInputFormat;
import org.apache.hadoop.mapreduce.lib.output.FileOutputFormat;

public class LogCleaner {
    public static class LogCleanerMapper extends Mapper<LongWritable, Text, Text, Text> {
        private Text ip = new Text();
        private Text log = new Text();

        public void map(LongWritable key, Text value, Context context) throws IOException, InterruptedException {
            String[] fields = value.toString().split(" ");
            ip.set(fields[0]);
            log.set(fields[1]);
            context.write(ip, log);
```

```java
        }
    }

    public static class LogCleanerReducer extends Reducer<Text, Text, Text, Text> {
        private Text result = new Text();

        public void reduce(Text key, Iterable<Text> values, Context context) throws IOException, InterruptedException {
            StringBuilder sb = new StringBuilder();
            Iterator<Text> iterator = values.iterator();
            while (iterator.hasNext()) {
                sb.append(iterator.next().toString());
                sb.append("\n");
            }
            result.set(sb.toString());
            context.write(key, result);
        }
    }

    public static void main(String[] args) throws Exception {
        Configuration conf = new Configuration();
        Job job = Job.getInstance(conf, "Log Cleaner");
        job.setJarByClass(LogCleaner.class);
        job.setMapperClass(LogCleanerMapper.class);
        job.setReducerClass(LogCleanerReducer.class);
        job.setOutputKeyClass(Text.class);
        job.setOutputValueClass(Text.class);
        FileInputFormat.addInputPath(job, new Path(args[0]));
        FileOutputFormat.setOutputPath(job, new Path(args[1]));
        System.exit(job.waitForCompletion(true) ? 0 : 1);
    }
}
```

在代码中定义了一个 LogCleanerMapper 类和一个 LogCleanerReducer 类。在 LogCleanerMapper 类中，我们将每个 IP 地址的日志作为值，将 IP 地址作为键，以便在 LogCleanerReducer 类中对每个 IP 地址的日志进行归约。在 LogCleanerReducer 类中，我们使用迭代器来将每个 IP 地址的日志连接成一个字符串，并将其作为值输出。

如果您想要模拟日志数据，可以创建一个文本文件，其中每一行都包含一个 IP 地址和一个日志条目，用空格分隔，如以下案例。

```
192.168.0.1 This is a log entry.
192.168.0.2 This is another log entry.
192.168.0.1 This is a third log entry.
```

▶▶ 2.10.2 统计电影最高评分

在大数据时代，电影评分数据广泛应用于电影推荐、趋势分析等领域。通过对海量电影评分数据进行统计和分析，可以挖掘出每部电影的最高评分，为电影消费者提供更有价值的参考信息。本次实战将利用 **MapReduce** 进行电影最高评分的统计，以分布式计算的方式对海量数据进行高效处理。

电影最高评分统计项目旨在从电影评分数据中提取每部电影的最高评分。使用 **MapReduce** 统计电影最高评分的功能，代码如下。

```java
import java.io.IOException;
import java.util.Iterator;
import java.util.Map;
import java.util.TreeMap;
import org.apache.hadoop.conf.Configuration;
import org.apache.hadoop.fs.Path;
import org.apache.hadoop.io.DoubleWritable;
import org.apache.hadoop.io.LongWritable;
import org.apache.hadoop.io.Text;
import org.apache.hadoop.mapreduce.Job;
import org.apache.hadoop.mapreduce.Mapper;
import org.apache.hadoop.mapreduce.Reducer;
import org.apache.hadoop.mapreduce.lib.input.FileInputFormat;
import org.apache.hadoop.mapreduce.lib.output.FileOutputFormat;

public class TopRatedMovies {
    public static class TopRatedMoviesMapper extends Mapper<LongWritable, Text, Text, DoubleWritable> {
        private Text movieId = new Text();
        private DoubleWritable rating = new DoubleWritable();

        public void map(LongWritable key, Text value, Context context) throws IOException, InterruptedException {
            String[] fields = value.toString().split(",");
            movieId.set(fields[1]);
            rating.set(Double.parseDouble(fields[2]));
            context.write(movieId, rating);
        }
    }

    public static class TopRatedMoviesReducer extends Reducer<Text, DoubleWritable, Text, DoubleWritable> {
        private TreeMap<Double, Text> topMovies = new TreeMap<Double, Text>();

        public void reduce (Text key, Iterable<DoubleWritable> values, Context context) throws IOException, InterruptedException {
            double sum = 0;
            int count = 0;
            for (DoubleWritable value : values) {
                sum += value.get();
                count++;
```

```java
            }
            double avgRating = sum / count;
            topMovies.put(avgRating, new Text(key));
            if (topMovies.size() > 10) {
                topMovies.remove(topMovies.firstKey());
            }
        }

        protected void cleanup(Context context) throws IOException, InterruptedException {
            for (Map.Entry<Double, Text> entry : topMovies.descendingMap().entrySet()) {
                context.write(entry.getValue(), new DoubleWritable(entry.getKey()));
            }
        }
    }

    public static void main(String[] args) throws Exception {
        Configuration conf = new Configuration();
        Job job = Job.getInstance(conf, "Top Rated Movies");
        job.setJarByClass(TopRatedMovies.class);
        job.setMapperClass(TopRatedMoviesMapper.class);
        job.setReducerClass(TopRatedMoviesReducer.class);
        job.setOutputKeyClass(Text.class);
        job.setOutputValueClass(DoubleWritable.class);
        FileInputFormat.addInputPath(job, new Path(args[0]));
        FileOutputFormat.setOutputPath(job, new Path(args[1]));
        System.exit(job.waitForCompletion(true) ? 0 : 1);
    }
}
```

在代码中，定义了一个 TopRatedMoviesMapper 类和一个 TopRatedMoviesReducer 类。在 TopRatedMoviesMapper 类中，我们将每个电影的评分作为值，将电影 ID 作为键，以便在 TopRatedMoviesReducer 类中对每个电影进行归约。在 TopRatedMoviesReducer 类中，我们使用 TreeMap 来存储电影和它们的平均评分，并在 cleanup 方法中将前 10 个最高评分的电影输出。

如果您想要模拟电影数据，可以创建一个 CSV 文件，其中每一行都包含一个用户 ID、电影 ID 和评分，用逗号分隔，如以下案例。

```
1,1,5.0
2,2,4.5
3,1,4.0
4,3,3.5
5,1,5.0
6,2,4.0
7,1,4.5
8,3,3.0
9,1,5.0
10,2,4.5
```

第3章 分布式协调服务 Zookeeper

本章内容介绍 ZooKeeper 的基本概念和工作原理，包括分布式协调、数据管理、状态同步等。学习 ZooKeeper 的安装和配置，以及集群模式的部署。掌握 ZooKeeper 的常用命令和操作，包括节点的创建、读取、更新和删除等操作。实践 ZooKeeper 的应用案例，包括分布式系统、数据监听等场景。

3.1 ZooKeeper 简介

Zookeeper 是一种分布式协调服务，它可以用于协调分布式系统中的各个节点。Zookeeper 提供了一个类似于文件系统的数据结构，可以用于存储和管理分布式系统中的配置信息、状态信息等。Zookeeper 还提供了一些 API，可以用于实现分布式锁、分布式队列等功能。

Zookeeper 特点如下。

1）高可用性：Zookeeper 采用了主从架构，可以保证在主节点宕机的情况下，集群仍然可以正常工作。

2）数据一致性：Zookeeper 使用了 ZAB 协议（Zookeeper Atomic Broadcast），可以保证数据在集群中的一致性。

3）实时性：Zookeeper 可以实时地响应客户端请求，可以用于实现分布式锁等功能。

4）可扩展性：Zookeeper 可以通过添加节点来扩展集群的规模。

3.2 ZooKeeper 结构和工作原理

简单来说 ZooKeeper 就是：文件系统 + 监听机制。

ZooKeeper 的核心是原子广播，这个机制保证了各个 Server 之间的同步。实现这个机制的协议叫作 ZAB 协议。

ZAB 协议有两种模式，它们分别是选主模式和广播模式。当服务启动或者在领导者崩溃后，ZAB 就进入了选主模式，当领导者被选举出来，且大多数 Server 完成了和 Leader 的状态同步以后，选主模式就结束了。

3.2.1 ZooKeeper 集群角色

在 ZooKeeper 集群当中，服务器角色有 Leader、Follower 和 Observer，如图 3-1 所示。

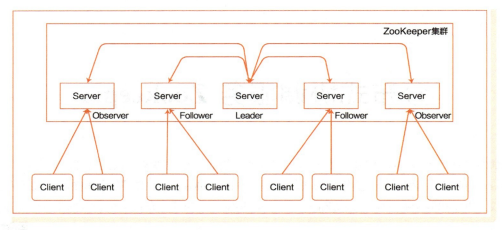

● 图 3-1　ZooKeeper 结构图

ZooKeeper 集群角色以及它们的功能

1) Leader：Zookeeper 集群中的 Leader 是一个主节点，它负责协调集群中的所有操作。Leader 节点负责处理客户端请求、维护集群状态、进行数据同步等操作。如果 Leader 节点宕机，集群会重新选举一个新的 Leader 节点。

2) Follower：Zookeeper 集群中的 Follower 节点是普通节点，它们负责接收客户端请求并将请求转发给 Leader 节点。Follower 节点还负责从 Leader 节点同步数据，以保证集群中的数据一致性。

3) Observer：Zookeeper 集群中的 Observer 节点也是普通节点，它们类似于 Follower 节点，但是不参与 Leader 选举过程。Observer 节点可以接收客户端请求并将请求转发给 Leader 节点，但是它们不参与数据同步过程。Observer 节点可以提高集群的读取性能，因为它们不需要参与数据同步，所以可以减轻 Leader 节点的负担。

总之，Zookeeper 集群中的 Leader 节点负责协调集群中的所有操作，Follower 节点负责接收客户端请求并将请求转发给 Leader 节点，Observer 节点类似于 Follower 节点，但是不参与数据同步过程。这些角色共同协作，保证了 Zookeeper 集群的高可用性和数据一致性。

3.2.2 ZooKeeper 的数据结构

在 ZooKeeper 中，数据结构的设计和实现是 ZooKeeper 的核心之一。通过了解 ZooKeeper 的数据结构，我们可以更好地理解和使用 ZooKeeper，并确保分布式系统的可靠性和一致性。

ZooKeeper 会维护一个具有层次关系的数据结构 Znode，它非常类似于一个标准的文件系统，如图 3-2 所示。接下来我们深入探讨 ZooKeeper 的数据结构，并通过实际操作来掌握如何使用 ZooKeeper 进行数据管理和协调。

1. ZooKeeper 基本概念

1）节点（Node）：ZooKeeper 中的最小数据单元，类似于文件系统中的文件或目录。

2）路径（Path）：节点在 ZooKeeper 中的唯一标识符，类似于文件系统中的路径。

3）版本（Version）：每个节点都有一个版本号，用于标识节点的变更历史。

4）会话（Session）：客户端与 ZooKeeper 服务器之间的连接，用于发送请求和接收响应。

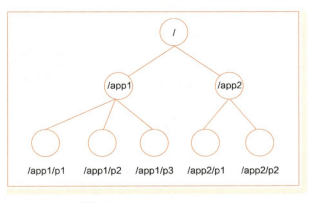

● 图 3-2 ZooKeeper 数据结构图

5）监控（Watcher）：ZooKeeper 的 Watch 机制是一种事件机制，用于监视 ZooKeeper 节点的状态变化。当一个节点的状态发生变化时，ZooKeeper 会通知所有注册了该节点 Watch 的客户端。

2. Znode 节点类型

Znode 分为临时节点（Ephemeral Node）和永久节点（Persistent Node）。节点的类型在创建时即被确定，并且不能被改变。

1）持久节点：创建后一直存在，直到被显式删除。可以包含数据和子节点。

2）临时节点：只在创建它们的会话期间存在，会话结束后自动删除。可以包含数据和子节点。

3）持久顺序节点：创建后一直存在，直到被显式删除。每个节点都有一个唯一的顺序号，可以用于排序。可以包含数据和子节点。

4）临时顺序节点：只在创建它们的会话期间存在，会话结束后自动删除。每个节点都有一个唯一的顺序号，可以用于排序。可以包含数据和子节点。

3. Znode 的监视（Watch）

ZooKeeper 支持 Watch，客户端可以在 Znode 上设置 Watch。Znode 更改时，将触发并删除监视。触发监视后，客户端会收到一个数据包，说明 Znode 已更改。如果客户端与其中一个 ZooKeeper 服务器之间的连接断开，则客户端将收到本地通知。

▶▶ 3.2.3　ZooKeeper 的工作流程

在 ZooKeeper 中，工作流程是指一系列的步骤和过程，用于实现分布式系统的协调和管理。通过了解 ZooKeeper 的工作流程，我们可以更好地理解和使用 ZooKeeper，并确保分布式系统的可靠性和高效性。ZooKeeper 的工作流程，如图 3-3 所示。

1. ZooKeeper 写入数据流程

1）客户端发送写入数据的请求，这个请求最终会被 Leader 处理。

2）Leader 会先写入数据，写入完成之后通知 Follower 进行数据的同步。

3）Follower 就会开始进行数据的同步（并行，多台 Follower 并行同步）。

4）每一个 Follower 只要数据同步完成，就会向 Leader 发送数据同步成功信息。

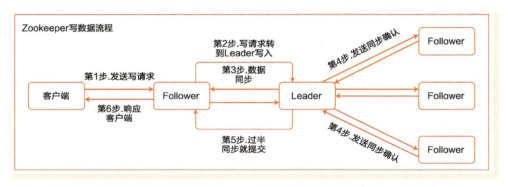

• 图 3-3 ZooKeeper 工作流程图

5）Leader 接收到超过半数以上的成功信息后，则认为这次写数据成功，其他节点慢慢进行同步，在数据同步的过程中，不对外提供读写服务。

6）响应客户端写入成功。

2. Leader 工作流程

1）ZooKeeper 集群中的所有节点都可以做 Leader，但只有一个节点可以最终成为 Leader。

2）当一个节点想要成为 Leader 时，它会向集群中的其他节点发送投票请求。

3）如果超过半数的节点同意该节点成为 Leader，则该节点成为 Leader。

4）如果某个节点成为 Leader 后，它会向集群中的其他节点发送心跳消息，以保持自己的 Leader 地位。

5）如果 Leader 节点失效或无法与其他节点通信，则集群中的其他节点会重新进行 Leader 选举。

总之，ZooKeeper 的 Leader 工作流程是通过投票和心跳消息来选举和维护 Leader 节点，以保证集群的正常运行。

3. Follower 工作流程

1）Follower 节点连接到 ZooKeeper 集群，并创建一个会话。

2）Follower 节点会向 Leader 节点发送心跳消息，以表明自己的存在。

3）Follower 节点会接收来自 Leader 节点的更新消息，并将其应用到本地数据存储中。

4）Follower 节点可以处理客户端的读取请求，但不能处理写入请求。

5）如果 Follower 节点在一段时间内没有收到来自 Leader 节点的心跳消息，则认为 Leader 节点失效，会重新进行 Leader 选举。

总之，ZooKeeper 的 Follower 工作流程是通过接收来自 Leader 节点的更新消息，并向 Leader 节点发送心跳消息来维护集群的正常运行。如果 Leader 节点失效，则 Follower 节点会重新进行 Leader 选举。

3.2.4 ZooKeeper 的监听器

在 ZooKeeper 中，监听器是一种机制，用于监控节点变化并触发相应的处理逻辑。通过使用监听器，我们可以实现分布式系统中数据的实时响应和处理。

1. 监听器概念

首先，我们需要了解 ZooKeeper 监听器的概念和作用。监听器是一种注册机制，客户端可以在节点上注册监听器，当节点数据发生变化时，监听器会自动触发相应的处理逻辑。了解监听器的概念和原理，将帮助你更好地理解 ZooKeeper 的实时性和可靠性。

2. 监听原理

当监听器被触发时，客户端可以执行相应的处理逻辑。处理逻辑可以是更新本地数据、发送消息给其他客户端等。了解监听器处理逻辑的编写，将帮助你更好地理解 ZooKeeper 的分布式协调和管理功能。ZooKeeper 的监听器原理，如图 3-4 所示。

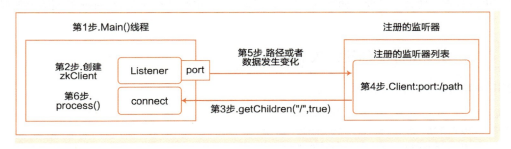

● 图 3-4　ZooKeeper 监听器原理

3. 监听流程

1）首先要有一个 main( ) 线程。

2）在 main 线程中创建 ZooKeeper 客户端，这时会创建两个线程，一个负责网络连接通信（connect），一个负责监听（listener）。

3）通过 connect 线程将注册的监听事件发送给 ZooKeeper。

4）在 ZooKeeper 的注册监听器列表中将注册的监听事件添加到列表中。

5）ZooKeeper 监听到有数据或路径发生变化时，就会将这个消息发送给 listener 线程。

6）在 listener 线程内部调用 process( ) 方法。

4. Watch 类型

1）监听节点数据的变化，getData 和 exists：返回关于节点的数据信息。

2）监听子节点增减的变化，getChildren：返回子节点列表。

5. Watch 注册与触发

在 ZooKeeper 中，客户端可以通过调用 exists 或 get 等方法来注册监听器。当节点数据发生变化时，ZooKeeper 会自动通知相应的监听器。掌握监听器的注册方式将帮助你更好地理解 ZooKeeper 的监听器使用和管理。

1）exists 操作上的 Watch，在被监视的 Znode 创建、删除或数据更新时被触发。

2）getData 操作上的 Watch，在被监视的 Znode 删除或数据更新时被触发。在被创建时不能被触发，因为只有 Znode 一定存在，getData 操作才会成功。

3）getChildren 操作上的 Watch，在被监视的 Znode 的子节点创建或删除，或是这个 Znode 自身被删除时被触发。可以通过查看 Watch 事件类型来区分是 Znode 还是它的子节点被删除：NodeDelete 表示 Znode 被删除，NodeDeletedChanged 表示子节点被删除。

3.3 ZooKeeper 实战

最后，我们将通过一些具体的实战操作来展示如何使用 ZooKeeper 的 API 操作，以及 ZooKeeper 监听器机制进行分布式协调和管理。通过这些实战操作，你将更好地理解 ZooKeeper 的监听器机制和实际应用。

3.3.1 ZooKeeper 创建持久节点

使用 ZooKeeper 创建持久节点的代码如下。

```java
import org.apache.zookeeper.CreateMode;
import org.apache.zookeeper.ZooDefs;
import org.apache.zookeeper.ZooKeeper;
import org.apache.zookeeper.data.ACL;
import org.apache.zookeeper.data.Id;
import org.apache.zookeeper.ZooKeeper;
import java.util.ArrayList;
import java.util.List;

public class ZooKeeperCreateNodeExample {

    public static void main(String[] args) throws Exception {
        String hostPort = "localhost:2181";
        int sessionTimeout = 5000;
        ZooKeeper zk = new ZooKeeper(hostPort, sessionTimeout, null);
        String path = "/myNode";
        byte[] data = "myData".getBytes();
        List<ACL> acl = new ArrayList<ACL>();
        Id id = new Id("world", "anyone");
        acl.add(new ACL(ZooDefs.Perms.ALL, id));
        CreateMode createMode = CreateMode.PERSISTENT;
        zk.create(path, data, acl, createMode);
        zk.close();
    }
}
```

代码中首先创建了一个 ZooKeeper 对象，用于连接到 ZooKeeper 服务器。然后，我们定义了要创建的节点的路径、数据、访问控制列表（ACL）和创建模式。最后，使用 ZooKeeper 对象的 create 方法创建了一个持久节点，并在创建节点时指定了路径、数据、ACL 和创建模式。

3.3.2 ZooKeeper 创建临时节点

使用 ZooKeeper 创建临时节点的代码如下。

```java
import org.apache.zookeeper.CreateMode;
import org.apache.zookeeper.ZooDefs;
import org.apache.zookeeper.ZooKeeper;
import org.apache.zookeeper.data.ACL;
import org.apache.zookeeper.data.Id;
import org.apache.zookeeper.ZooKeeper;
import java.util.ArrayList;
import java.util.List;

public class ZooKeeperCreateEphemeralNodeExample {
    public static void main(String[] args) throws Exception {
        String hostPort = "localhost:2181";
        int sessionTimeout = 5000;
        ZooKeeper zk = new ZooKeeper(hostPort, sessionTimeout, null);
        String path = "/myEphemeralNode";
        byte[] data = "myData".getBytes();
        List<ACL> acl = new ArrayList<ACL>();
        Id id = new Id("world", "anyone");
        acl.add(new ACL(ZooDefs.Perms.ALL, id));
        CreateMode createMode = CreateMode.EPHEMERAL;
        zk.create(path, data, acl, createMode);
        zk.close();
    }
}
```

代码中首先创建了一个 ZooKeeper 对象，用于连接到 ZooKeeper 服务器。然后，我们定义了要创建的节点的路径、数据、访问控制列表（ACL）和创建模式。最后，使用 ZooKeeper 对象的 create 方法创建了一个临时节点，并在创建节点时指定了路径、数据、ACL 和创建模式。

3.3.3 ZooKeeper 递归创建节点

使用 ZooKeeper 递归创建节点的代码如下。

```java
import org.apache.zookeeper.CreateMode;
import org.apache.zookeeper.ZooDefs;
import org.apache.zookeeper.ZooKeeper;
import org.apache.zookeeper.data.ACL;
import org.apache.zookeeper.data.Id;
import org.apache.zookeeper.ZooKeeper;
import java.util.ArrayList;
import java.util.List;
```

```java
public class ZooKeeperCreateRecursiveNodeExample {
    public static void main(String[] args) throws Exception {
        String hostPort = "localhost:2181";
        int sessionTimeout = 5000;
        ZooKeeper zk = new ZooKeeper(hostPort, sessionTimeout, null);
        String path = "/my/recursive/node";
        byte[] data = "myData".getBytes();
        List<ACL> acl = new ArrayList<ACL>();
        Id id = new Id("world", "anyone");
        acl.add(new ACL(ZooDefs.Perms.ALL, id));
        CreateMode createMode = CreateMode.PERSISTENT;
        zk.create(path, data, acl, createMode, true);
        zk.close();
    }
}
```

在这个代码中，我们首先创建了一个 ZooKeeper 对象，用于连接到 ZooKeeper 服务器。然后，定义了要创建的节点的路径、数据、访问控制列表（ACL）和创建模式。最后，使用 ZooKeeper 对象的 create 方法创建了一个递归节点，并在创建节点时指定了路径、数据、ACL、创建模式和 createParents 参数。

3.3.4 ZooKeeper 读取数据

使用 ZooKeeper 读取数据的代码如下。

```java
import org.apache.zookeeper.ZooKeeper;
import org.apache.zookeeper.data.Stat;

public class ZooKeeperGetDataExample {
    public static void main(String[] args) throws Exception {
        String hostPort = "localhost:2181";
        int sessionTimeout = 5000;
        ZooKeeper zk = new ZooKeeper(hostPort, sessionTimeout, null);
        String path = "/myNode";
        Stat stat = new Stat();
        byte[] data = zk.getData(path, null, stat);
        String dataString = new String(data);
        System.out.println("Data: " + dataString);
        System.out.println("Version: " + stat.getVersion());
        zk.close();
    }
}
```

在这个代码中，我们首先创建了一个 ZooKeeper 对象，用于连接到 ZooKeeper 服务器。然后，定义了要读取的节点的路径，并使用 getData 方法读取了节点的数据。我们还使用 Stat 对象获取了节点的元数据，例如版本号。最后，将节点的数据打印到控制台上。

▶▶ 3.3.5 ZooKeeper 更新数据

使用 ZooKeeper 更新数据的代码如下。

```java
import org.apache.zookeeper.ZooKeeper;
import org.apache.zookeeper.data.Stat;

public class ZooKeeperSetDataExample {

    public static void main(String[] args) throws Exception {
        String hostPort = "localhost:2181";
        int sessionTimeout = 5000;
        ZooKeeper zk = new ZooKeeper(hostPort, sessionTimeout, null);
        String path = "/myNode";
        byte[] data = "newData".getBytes();
        Stat stat = zk.setData(path, data, -1);
        System.out.println("Version: " + stat.getVersion());
        zk.close();
    }
}
```

在这个代码中，我们首先创建了一个 ZooKeeper 对象，用于连接到 ZooKeeper 服务器。然后，定义了要更新的节点的路径，并使用 setData 方法更新了节点的数据。我们还使用 Stat 对象获取了节点的元数据，例如版本号。最后，将节点的版本号打印到控制台上。

▶▶ 3.3.6 ZooKeeper 监听节点

使用 ZooKeeper 监听节点的代码如下。

```java
import org.apache.zookeeper.*;
import org.apache.zookeeper.data.Stat;

import java.io.IOException;
import java.util.concurrent.CountDownLatch;

public class ZooKeeperWatchExample implements Watcher {

    private static final int SESSION_TIMEOUT = 5000;
    private static final String HOST_PORT = "localhost:2181";
    private static final String PATH = "/myNode";

    private ZooKeeper zk;
    private CountDownLatch connectedSignal = new CountDownLatch(1);

    public void connect() throws IOException, InterruptedException {
```

```java
        zk = new ZooKeeper(HOST_PORT, SESSION_TIMEOUT, this);
        connectedSignal.await();
    }

    public void close() throws InterruptedException {
        zk.close();
    }

    public void process(WatchedEvent event) {
        if (event.getState() == Event.KeeperState.SyncConnected) {
            connectedSignal.countDown();
        }
        if (event.getType() == Event.EventType.NodeDataChanged) {
            try {
                byte[] data = zk.getData(PATH, this, null);
                String dataString = new String(data);
                System.out.println("节点数据变化:" + dataString);
            } catch (KeeperException e) {
                e.printStackTrace();
            } catch (InterruptedException e) {
                e.printStackTrace();
            }
        }
    }

    public void watchNode() throws KeeperException, InterruptedException {
        Stat stat = zk.exists(PATH, this);
        if (stat != null) {
            byte[] data = zk.getData(PATH, this, null);
            String dataString = new String(data);
            System.out.println("节点数据为:" + dataString);
        } else {
            System.out.println("节点不存在");
        }
    }

    public static void main(String[] args) throws Exception {
        ZooKeeperWatchExample example = new ZooKeeperWatchExample();
        example.connect();
        example.watchNode();
        Thread.sleep(Long.MAX_VALUE);
        example.close();
    }
}
```

在代码中，我们首先创建了一个 ZooKeeper 对象，用于连接到 ZooKeeper 服务器。然后，定义了要监听的节点的路径，并使用 exists 方法检查节点是否存在。如果节点存在，使用 getData 方法获取节点的数据，并将其打印到控制台上。如果节点不存在，我们将"节点不存在"打印到控制台上。

接下来，使用 Watcher 接口实现了一个 process 方法，用于处理节点变化事件。在 process 方法中，我们首先检查事件的类型是否为 NodeDataChanged，如果是，使用 getData 方法获取节点的数据，并将其打印到控制台上。

最后，我们在 main 方法中创建了一个 ZooKeeperWatchExample 对象，并调用了 connect 方法连接到 ZooKeeper 服务器。然后，调用了 watchNode 方法，用于监听节点的变化。最后，使用 Thread.sleep(Long.MAX_VALUE) 方法使程序保持运行状态，直到手动停止它。

▶▶ 3.3.7　ZooKeeper 监听子节点

使用 ZooKeeper 监听子节点的代码如下。

```java
import org.apache.zookeeper.*;
import org.apache.zookeeper.data.Stat;
import java.io.IOException;
import java.util.List;
import java.util.concurrent.CountDownLatch;

public class ZooKeeperWatchExample implements Watcher {

    private static final int SESSION_TIMEOUT = 5000;
    private static final String HOST_PORT = "localhost:2181";
    private static final String PATH = "/myNode";

    private ZooKeeper zk;
    private CountDownLatch connectedSignal = new CountDownLatch(1);

    public void connect() throws IOException, InterruptedException {
        zk = new ZooKeeper(HOST_PORT, SESSION_TIMEOUT, this);
        connectedSignal.await();
    }

    public void close() throws InterruptedException {
        zk.close();
    }

    public void process(WatchedEvent event) {
        if (event.getState() == Event.KeeperState.SyncConnected) {
            connectedSignal.countDown();
        }
        if (event.getType() == Event.EventType.NodeChildrenChanged) {
            try {
                List<String> children = zk.getChildren(PATH, this);
                System.out.println("子节点变化:" + children);
```

```java
        } catch (KeeperException e) {
            e.printStackTrace();
        } catch (InterruptedException e) {
            e.printStackTrace();
        }
    }
}

public void watchChildren() throws KeeperException, InterruptedException {
    List<String> children = zk.getChildren(PATH, this);
    System.out.println("子节点为:" + children);
}

public static void main(String[] args) throws Exception {
    ZooKeeperWatchExample example = new ZooKeeperWatchExample();
    example.connect();
    example.watchChildren();
    Thread.sleep(Long.MAX_VALUE);
    example.close();
}
}
```

代码中,我们首先创建了一个 ZooKeeper 对象,用于连接到 ZooKeeper 服务器。然后,定义了要监听的子节点的路径,并使用 getChildren 方法检查子节点是否存在。如果子节点存在,使用 getChildren 方法获取子节点的列表,并将其打印到控制台上。如果子节点不存在,将"子节点不存在"打印到控制台上。

接下来,使用 Watcher 接口实现了一个 process 方法,用于处理子节点变化事件。在 process 方法中,我们首先检查事件的类型是否为 NodeChildrenChanged,如果是,使用 getChildren 方法获取子节点的列表,并将其打印到控制台上。

最后,我们在 main 方法中创建了一个 ZooKeeperWatchExample 对象,并调用了 connect 方法连接到 ZooKeeper 服务器。然后,调用了 watchChildren 方法,用于监听子节点的变化。最后,使用 Thread.sleep(Long.MAX_VALUE) 方法使程序保持运行状态,直到手动停止它。

▶▶ 3.3.8 ZooKeeper 实现服务注册与发现

在分布式系统中,服务注册是指将服务的信息发布到系统中,以便其他服务或客户端能够发现和访问该服务。服务发现则是从系统中获取服务的信息,以便客户端能够找到所需的服务。ZooKeeper 通过提供一个集中式的服务注册和发现机制,使得分布式系统中的服务能够动态地注册和发现彼此。使用 ZooKeeper 实现服务注册的代码如下。

```java
package com.jareny.bigdata.zookeeper;

import org.apache.curator.framework.CuratorFramework;
import org.apache.curator.framework.CuratorFrameworkFactory;
```

```java
import org.apache.curator.framework.recipes.cache.*;
import org.apache.curator.retry.RetryNTimes;
import org.apache.zookeeper.CreateMode;
import org.springframework.beans.factory.annotation.Autowired;
import org.springframework.boot.SpringApplication;
import org.springframework.boot.autoconfigure.SpringBootApplication;
import org.springframework.context.annotation.Bean;
import org.springframework.web.bind.annotation.GetMapping;
import org.springframework.web.bind.annotation.RestController;
import java.util.List;
import static org.apache.curator.framework.recipes.cache.PathChildrenCacheEvent.Type.*;

@SpringBootApplication
@RestController
public class ServerApplication {
    private static final String PATH = "/example/server";
    private static final String SERVER_PATH = "/example/server/s";

    @Autowired
    private CuratorFramework client;

    @Bean(initMethod = "start")
    public NodeCache nodeCache() {
        NodeCache nodeCache = new NodeCache(client, SERVER_PATH);
        try {
            nodeCache.start();
        } catch (Exception e) {
            throw new RuntimeException(e);
        }
        nodeCache.getListenable().addListener(new NodeCacheListener() {
            @Override
            public void nodeChanged() throws Exception {
                ChildData data = nodeCache.getCurrentData();
                if (data == null) {
                    System.out.println("Server node deleted!");
                } else {
                    System.out.println("Server node changed: " + new String(data.getData()));
                }
            }
        });
        return nodeCache;
    }

    @Bean(initMethod = "start")
    public PathChildrenCache pathChildrenCache() {
        PathChildrenCache pathChildrenCache = new PathChildrenCache(client, PATH, true);
```

```java
        try {
            pathChildrenCache.start(PathChildrenCache.StartMode.POST_INITIALIZED_EVENT);
        } catch (Exception e) {
            throw new RuntimeException(e);
        }
        pathChildrenCache.getListenable().addListener(new PathChildrenCacheListener() {
            @Override
            public void childEvent(CuratorFramework client, PathChildrenCacheEvent event) throws Exception {
                PathChildrenCacheEvent.Type type = event.getType();
                if (type == CHILD_ADDED) {
                    System.out.println("Server added: " + event.getData().getPath());
                } else if (type == CHILD_REMOVED) {
                    System.out.println("Server removed: " + event.getData().getPath());
                } else if (type == CHILD_UPDATED) {
                    System.out.println("Server updated: " + event.getData().getPath());
                }
            }
        });
        return pathChildrenCache;
    }

    @GetMapping("/servers")
    public List<String> servers() throws Exception {
        return client.getChildren().forPath(PATH);
    }

    public static void main(String[] args) {
        SpringApplication.run(ServerApplication.class, args);
    }

    @Bean(initMethod = "start")
    public CuratorFramework curatorFramework() {
        CuratorFramework client = CuratorFrameworkFactory.newClient("192.168.81.111:2181", new RetryNTimes(5, 1000));
        client.start();
        try {
client.create().creatingParentsIfNeeded().withMode(CreateMode.EPHEMERAL_SEQUENTIAL).forPath(SERVER_PATH, "server".getBytes());
        } catch (Exception e) {
            throw new RuntimeException(e);
        }
        return client;
    }
}
```

使用 ZooKeeper 实现服务发现的代码如下。

```java
package com.jareny.bigdata.zookeepr;

import org.apache.curator.framework.CuratorFramework;
import org.apache.curator.framework.CuratorFrameworkFactory;
import org.apache.curator.framework.recipes.cache.*;
import org.apache.curator.retry.RetryNTimes;
import org.springframework.beans.factory.annotation.Autowired;
import org.springframework.boot.SpringApplication;
import org.springframework.boot.autoconfigure.SpringBootApplication;
import org.springframework.context.annotation.Bean;
import org.springframework.web.bind.annotation.GetMapping;
import org.springframework.web.bind.annotation.RestController;
import java.util.List;
import static org.apache.curator.framework.recipes.cache.PathChildrenCacheEvent.Type.*;

@SpringBootApplication
@RestController
public class ClientApplication {
    private static final String PATH = "/example/server";

    @Autowired
    private CuratorFramework client;

    @Bean(initMethod = "start")
    public NodeCache nodeCache() {
        NodeCache nodeCache = new NodeCache(client, PATH);
        try {
            nodeCache.start();
        } catch (Exception e) {
            throw new RuntimeException(e);
        }
        nodeCache.getListenable().addListener(new NodeCacheListener() {
            @Override
            public void nodeChanged() throws Exception {
                ChildData data = nodeCache.getCurrentData();
                if (data == null) {
                    System.out.println("Server node deleted!");
                } else {
                    System.out.println("Server node changed: " + new String(data.getData()));
                }
            }
        });
        return nodeCache;
    }
}
```

```java
    @Bean(initMethod = "start")
    public PathChildrenCache pathChildrenCache() {
        PathChildrenCache pathChildrenCache = new PathChildrenCache(client, PATH, true);
        try {
            pathChildrenCache.start(PathChildrenCache.StartMode.POST_INITIALIZED_EVENT);
        } catch (Exception e) {
            throw new RuntimeException(e);
        }
        pathChildrenCache.getListenable().addListener(new PathChildrenCacheListener() {
            @Override
             public void childEvent(CuratorFramework client, PathChildrenCacheEvent event) throws Exception {
                PathChildrenCacheEvent.Type type = event.getType();
                if (type == CHILD_ADDED) {
                    System.out.println("Server added: " + event.getData().getPath());
                } else if (type == CHILD_REMOVED) {
                    System.out.println("Server removed: " + event.getData().getPath());
                } else if (type == CHILD_UPDATED) {
                    System.out.println("Server updated: " + event.getData().getPath());
                }
            }
        });
        return pathChildrenCache;
    }

    @GetMapping("/servers")
    public List<String> servers() throws Exception {
        return client.getChildren().forPath(PATH);
    }

    public static void main(String[] args) {
        SpringApplication.run(ClientApplication.class, args);
    }

    @Bean(initMethod = "start")
    public CuratorFramework curatorFramework() {
        CuratorFramework client = CuratorFrameworkFactory.newClient("192.168.81.111:2181", new RetryNTimes(5, 1000));
        client.start();
        return client;
    }
}
```

第4章

数据仓库 Hive

本章内容介绍 Hive 的基本概念和工作原理，包括 Hive 的运行流程、HiveQL 语言、数据模型等。学习 Hive 的配置文件格式和各组件的作用。掌握 Hive 的常用命令和操作，包括创建表、加载数据、查询数据等操作。实践 Hive 的应用案例、数据分析等场景。

4.1 Hive 简介和特点

1. Hive 简介

Hive 是一个基于 Hadoop 的数据仓库工具，它提供了类似于 SQL 的查询语言，称为 HiveQL，可以用于分析大规模的数据集。Hive 将查询转换为 MapReduce 任务，并在 Hadoop 集群上执行这些任务。Hive 还提供了一些内置函数和 UDF（用户定义函数），可以用于处理和转换数据。

Hive 的数据模型类似于关系型数据库，它将数据存储在表中，并支持分区和桶（bucket）等概念。Hive 还支持外部表，可以将数据存储在 Hadoop 集群之外的存储系统中，如 HDFS、S3 等。

Hive 的优点包括易于使用、可扩展性强、支持复杂查询和数据处理等。但是，由于 Hive 是基于 MapReduce 的，所以它的查询性能相对较低，不适合实时查询和交互式分析。

2. Hive 特点

1）容易上手：可以将结构化数据映射成一张表，并且提供了类 SQL 查询语言 HQL。

2）海量数据的高性能查询和分析系统：Hive 的查询是通过 MapReduce 框架实现的，而 MapReduce 本身就是为实现针对海量数据的高性能处理而设计的。所以 Hive 天然就能高效地处理海量数据。

3）自定义函数：除了 HQL 自身提供的能力，用户还可以自定义其使用的数据类型，也可以用任何语言自定义 mapper 和 reducer 脚本，还可以自定义函数等。这就赋予了 HQL 极大的可扩展性。用户可以利用这种可扩展性实现非常复杂的查询。

4）高扩展性、容错性：Hive 本身并没有执行机制，用户查询的执行是通过 MapReduce 框架实现的。由于 MapReduce 框架本身具有高度可扩展和高容错的特点，所以 Hive 也相应具有这些特点。

5）与 Hadoop 其他组件完全兼容：Hive 自身并不存储数据，而是通过接口访问数据。这就使得 Hive 支持各种数据源和数据格式。Hive 支持处理 HDFS 上的多种文件格式，还支持处理 HBase 数据库。用户也完全可以实现自己的驱动来增加新的数据源和数据格式。

4.2 Hive 结构和原理

Hive 是 Apache 的一个开源项目，它提供了一个数据仓库基础设施，用于处理和查询大数据。Hive 的结构和原理使得它能够高效地处理大规模的数据，同时提供了类似于 SQL 的查询语言（HiveQL），使得对数据的查询和分析更加简便。

Hive 的结构主要包括以下几个部分：仓库（Repository）、元数据（Metadata）、驱动器（Driver）、查询语言（HiveQL）、执行引擎（Execution Engine）、用户接口（User Interface）。

Hive 的原理主要涉及以下几个过程：查询解析（Query Parsing）、查询计划（Query Planning）、执行计划（Query Execution）、结果返回（Result Return）。

4.2.1 Hive 结构

Hive 是一个构建在 Hadoop 上的数据仓库工具，通过了解 Hive 的架构组成，我们可以更好地理解和使用 Hive，并确保高效地进行数据分析和处理。

Hive 的架构组成包括：Hive Client、Hive Server、Metastore、Driver、Compiler 和 Execution Engine。它们分别负责用户接口、服务端、元数据存储、查询执行引擎等不同的功能，如图 4-1 所示。

Hive 的架构以及它们的作用如下。

1）Hive Client：Hive Client 是 Hive 的用户接口，它提供了类似于 SQL 的查询语言（HiveQL），用户可以通过 Hive Client 向 Hive Server 提交查询请求。

2）Hive Server：Hive Server 是 Hive 的服务端，它接收 Hive Client 提交的查询请求，并将请求转发给相应的组件进行处理。

3）Metastore：Metastore 是 Hive 的元数据存储组件，它存储了 Hive 表的元数据信息，如表名、列名、数据类型等。

4）Driver：Driver 是 Hive 的查询执

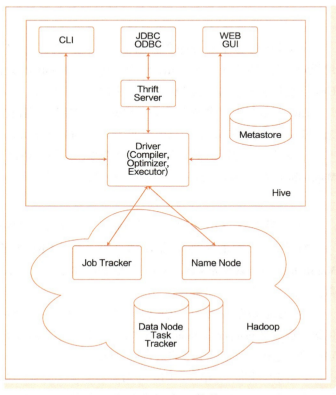

● 图 4-1 Hive 结构图

行引擎，它负责解析 HiveQL 语句，生成相应的执行计划，并将执行计划提交给相应的执行引擎进行处理。

5）Compiler：Compiler 是 Hive 的查询编译器，它将 HiveQL 语句编译成执行计划。

6）Execution Engine：Execution Engine 是 Hive 的查询执行引擎，它负责执行计划，并将结果返回给 Driver。

4.2.2 Hive 运行的流程

通过了解 Hive 的运行流程，我们可以更好地理解和使用 Hive，并确保高效地进行数据分析和处理。Hive 的运行流程，如图 4-2 所示。

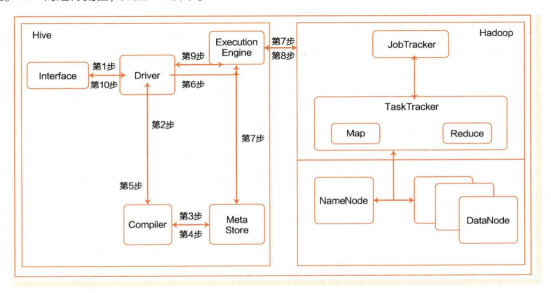

● 图 4-2 Hive 的运行流程

Hive 运行流程如下。

1）Execute Query：Hive 接口，如命令行或 Web UI 发送查询驱动程序（任何数据库驱动程序，如 JDBC、ODBC 等）来执行。

2）Get Plan：在驱动程序帮助下查询编译器，分析查询检查语法和查询计划或查询的要求。

3）Get Metadata：编译器发送元数据请求到 Metastore（任何数据库）。

4）Send Metadata：Metastore 接收到元数据请求后，会查询其存储的元数据，并将所需的元数据作为响应发送给编译器。

5）Send Plan：编译器检查要求，并重新发送计划给驱动程序。到此为止，查询解析和编译完成。

6）Execute Plan：驱动程序发送的执行计划到执行引擎。

7）Execute Job：在内部，执行引擎将执行计划转换为 MapReduce 作业，并提交给 Hadoop 集群执行。作业 Hadoop 的 JobTracker 和 TaskTracker 上运行，完成数据的处理和计算。在这里，查询执行

MapReduce工作。与此同时，在执行时，执行引擎可以通过 Metastore 执行元数据操作。

8）Fetch Result：执行引擎接收来自数据节点的结果。

9）Send Results：执行引擎发送这些结果值给驱动程序。

10）Send Results：驱动程序将结果发送给 Hive 接口。

4.2.3 Hive 的 HQL 转换过程

Hive 是 Apache Hadoop 生态系统中的一种数据仓库工具，它可以将结构化的数据文件映射为一张数据库表，并提供类似于 SQL 的查询功能。Hive 将 HQL（Hive Query Language）转换为 MapReduce 任务的过程可以分为以下几个步骤，如图 4-3 所示。

Hive 转换成 MapReduce 的具体过程如下。

1）由 Hive 驱动模块中的编译器对用户输入的 HQL 语言进行词法和语法解析，将 HQL 语句转换为抽象语法树的形式。

2）抽象语法树的结构仍很复杂，不方便直接翻译为 MapReduce 算法程序，因此，把抽象语法书转换为查询块。

3）把查询块转换成逻辑查询计划，里面包含了许多逻辑操作符。

4）重写逻辑查询计划，进行优化，合并多余操作，减少 MapReduce 任务数量。

5）将逻辑操作符转换成需要执行的具体 MapReduce 任务。

6）对生成的 MapReduce 任务进行优化，生成最终的 MapReduce 任务执行计划。

7）由 Hive 驱动模块中的执行器，对最终的 MapReduce 任务进行执行输出。

● 图 4-3 Hive 将 HQL 转换成 MapReduce 的过程

4.3 电商用户行为分析

4.3.1 项目背景及目的

1. 项目背景

移动互联网企业从粗放式到精细化运营管理过程中，需要结合市场、渠道、用户行为等数据分析，对用户开展有针对性的运营活动，提供个性化、差异化的运营策略，以实现运营业务指标。

本项目对电商用户行为数据进行分析，通过用户行为分析业务问题，提供针对性的运营策略。

2. 项目目标

本次分析的目标是通过用户行为数据分析，为以下问题提供解释和改进建议。

1）分析用户使用 App 过程中的常见电商分析指标，建立用户行为转化漏斗模型，确定各个环节的流失率，找到需要改进的环节。

2）研究用户在不同时间尺度下的行为规律，找到用户在不同时间周期下的活跃规律，在用户活跃时间点推出相应的营销策略。

3）找到用户对不同种类商品的偏好，找到针对不同商品的营销策略。

温馨提示：用于本项目的数据集文件 UserBehavior.txt 中存储练习数据集。可以从 https://gitee.com/jareny/jareny-bigdata.git 获取。

本数据集有行为的约一百万随机用户的所有行为（行为包括点击、购买、加购、喜欢）。由用户 ID、商品 ID、商品类目 ID、行为类型和时间戳组成，并以逗号分隔。关于数据集中每一列的数据结构详细描述，如表 4-1 所示。

表 4-1 商品数据采集表

用户 ID	商品 ID	商品类目 ID	行为类型	时间戳
1000	408860	551706	pv	1511543246
1000	4590199	4338287	pv	1511543742
1000	1976168	4338287	cart	1511543773
1000001	1649625	4145813	pv	1511802582
1000001	3093290	4145813	buy	1511571820
1000004	2980887	3607361	cart	1511669671
1000004	864562	4261392	pv	1511624303

4.3.2 数据导入

大数据 Hive 数据导入是指将大量数据从不同的数据源导入到 Hive 数据仓库中的过程。这个过程通常涉及以下步骤：数据准备、Hive 表创建、数据加载、数据转换、数据校验和监控。

总的来说，大数据 Hive 数据导入的过程需要考虑到数据的规模、格式、转换需求以及监控和校验的需求。通过合理的规划和实施，可以实现高效的数据导入和管理，为后续的数据分析和决策提供可靠的支持。

Hive 数据导入实现的步骤如下。

（1）连接 MySQL 并使用 jareny 数据库

```
hive> create database if not exists jarenyBigdata;
hive> use  jarenyBigdata;
OK
Time taken: 0.028 seconds
```

（2）创建新表 userbehavior

```
hive> create table userbehavior(user_id int,item_id int,category_id int,behavior string,time bigint) row format delimited fields terminated by '\t';
```

（3）载入源数据

```
hive> load data local inpath "/opt/jareny/bigdata/data/UserBehavior.txt" into table userbehavior;
```

4.3.3 数据清洗

大数据 Hive 数据清洗是指对导入到 Hive 数据仓库中的数据进行清洗和整理，以确保数据的准确性和完整性。这个过程通常涉及以下步骤：数据检查、空值处理、异常值处理、数据去重、数据转换、数据存储。

总的来说，大数据 Hive 数据清洗的过程需要考虑到数据的规模、格式、转换需求以及监控和校验的需求。通过合理的规划和实施，可以实现高质量的数据清洗和管理，为后续的数据分析和决策提供准确和完整的数据支持。

（1）删除重复值

```
hive> SELECT user_id,item_id,category_id,time FROM userbehavior GROUP BY user_id,item_id,
category_id,time HAVING count(user_id)>1;
```

（2）查看缺失值

```
hive>  SELECT count(user_id),count(item_id),count(category_id),count(behavior),count
(time) FROM userbehavior;
<!-- 省略运行的过程 -->
0 0 0 0 0
Time taken: 24.578 seconds, Fetched: 1 row(s)
```

（3）过滤异常值

1）由于数据集时间范围为 2017-11-25 至 2017-12-3，因此需要对不在该时间范围内的异常数据进行过滤。
2）筛选异常数据。

```
hive> SELECT * FROM userbehavior WHERE from_unixtime(time,'yyyy-MM-dd')<'2017-11-25' or from
_unixtime(time,'yyyy-MM-dd')>'2017-12-03';
<!-- 省略运行的过程,并且数据量过大,只显示部分数据-->
100249    4125603    2355072    pv    1511539191
1002591   3527580    582243     pv    1511528335
1002677   5001596    4220691    pv    1511520981
Time taken: 0.214 seconds, Fetched: 29 row(s)
```

3）过滤异常数据。

```
hive> DELETE FROM userbehavior WHERE WHERE from_unixtime(time,'yyyy-MM-dd')<'2017-11-25' or
from_unixtime(time,'yyyy-MM-dd')>'2017-12-03';
```

4.3.4 数据分析

1. 基于用户行为转化漏斗模型分析用户行为

漏斗分析模型已经广泛应用于各行业的数据分析工作中，用以评估总体转化率、各个环节的转化率，以科学评估促销专题活动效果等，通过与其他数据分析模型结合进行深度用户行为分析，从而找到用户流失的原因，以提升用户量、活跃度、留存率，并提升数据分析与决策的科学性等。

常用漏斗模型：首页—商品详情页—加入购物车—提交订单—支付订单。

本数据集只包含商品详情页（pv）、加入购物车（cart）、支付订单（buy）数据，因此将漏斗模型简化为：商品详情页—加入购物车—支付订单。

（1）用户总行为（PV）的转化漏斗

```
hive> SELECT behavior,COUNT(*) FROM userbehavior GROUP BY behavior order by behavior desc;
<!--省略运行的过程-->
pv      58892
fav     1648
cart    3598
buy     1398
Time taken: 52.368 seconds, Fetched: 4 row(s)
```

（2）独立访客转化漏斗

```
hive> SELECT behavior,count(DISTINCT user_id) FROM userbehavior GROUP BY behavior ORDER BY behavior DESC;
<!--省略运行的过程-->
pv      614
fav     235
cart    452
buy     423
Time taken: 51.642 seconds, Fetched: 4 row(s)
```

2. 用户行为转化漏斗模型分析

（1）每天的用户行为分析

```
hive>SELECT from_unixtime(time,'yyyy-MM-dd HH:mm:ss') as time,
count(user_id) ct,sum(if(behavior='pv',1,0)) pv,
sum(if(behavior='cart',1,0)) cart,
sum(if(behavior='fav',1,0)) fav,
sum(if(behavior='buy',1,0)) buy
FROM userbehavior
GROUP BY time;
<!--省略运行的过程-->
2017-12-03 23:57:33    1    1    0    0    0
2017-12-03 23:57:43    1    1    0    0    0
2017-12-03 23:57:50    1    1    0    0    0
2017-12-03 23:58:00    1    1    0    0    0
2017-12-03 23:58:37    1    1    0    0    0
2017-12-03 23:58:53    1    1    0    0    0
2017-12-03 23:58:54    1    1    0    0    0
2017-12-03 23:59:00    1    1    0    0    0
2017-12-03 23:59:02    1    1    0    0    0
Time taken: 30.445 seconds, Fetched: 61804 row(s)
```

（2）商品销量排行榜前 10

```
hive> SELECT item_id, count(behavior) as ct FROM userbehavior WHERE behavior='buy' GROUP BY item_id ORDER BY ct DESC limit 10;
```

```
<!-- 省略运行的过程 -->
1131455    3
1910706    3
551325     2
37314      2
642571     2
Time taken: 44.574 seconds, Fetched: 10 row(s)
```

(3) 商品浏览量排行榜前 10

```
hive> SELECT item_id, count(behavior) as ct FROM userbehavior WHERE behavior='pv' GROUP BY item_id  ORDER BY ct DESC limit 10;
<!-- 省略运行的过程 -->
3006793    31
3769601    27
2546537    26
812879     25
138964     23
1325619    22
4350284    22
2331370    21
3557403    21
1420923    20
Time taken: 79.492 seconds, Fetched: 10 row(s)
```

(4) 商品 Top10 所属的分类

```
hive> SELECT category_id
from (SELECT item_id,count(*) as ct from userbehavior group by item_id order by ct  limit 10)  t1
left join userbehavior u  on t1.item_id = u.item_id group by category_id limit 10;
<!-- 省略运行的过程 -->
965809
1029459
1080785
1813610
2195789
2920476
4048584
4357323
```

第 5 章

面向列的数据库 HBase

本章内容介绍 HBase 的基本概念和工作原理，包括 HBase 数据模型、HBase 架构、HBase 读写流程等。学习 HBase 的安装和配置，集群模式的部署。掌握 HBase 的常用命令和操作，包括创建表、插入数据、查询数据等操作。实践 HBase 的应用案例，包括数据存储、数据分析等场景。

5.1 HBase 简介

1. Hbase 概述

HBase 是一个分布式的、面向列的 NoSQL 数据库。它是基于 Hadoop 的 HDFS 文件系统构建的，可以处理大量的结构化和半结构化数据。

HBase 被设计用来提供高可靠性、高性能、列存储、可伸缩、多版本的 NoSQL 的分布式数据存储系统，实现对大型数据的实时、随机的读写访问。

2. Hbase 的特点

1）面向列：HBase 是一个面向列的数据库，它可以存储和检索大量的列数据。这使得 HBase 非常适合存储结构化和半结构化数据，例如日志、用户配置文件等。

2）分布式：HBase 是一个分布式的数据库，它可以在多个节点上存储数据。这使得 HBase 非常适合处理大量的数据，因为它可以水平扩展。

3）可扩展性：HBase 是一个可扩展的数据库，它可以根据需要添加更多的节点。这使得 HBase 非常适合处理大量的数据，因为它可以随着数据量的增加而扩展。

4）高可用性：HBase 是一个高可用的数据库，它可以在节点故障时自动恢复。这使得 HBase 非常适合处理关键业务数据，因为它可以保证数据的可靠性和可用性。

5）快速：HBase 是一个快速的数据库，它可以在毫秒级别内检索数据。这使得 HBase 非常适合处理实时数据，例如日志、传感器数据等。

5.2 HBase 架构

HBase 是一个构建在 Hadoop 上的分布式、可扩展、面向列的数据库，它提供了一种高效的方式来存储和管理大规模的数据。HBase 的集群架构为主从架构,主要由 HMaster、HRegionServer 和 ZooKeeper 组成。HBase 将逻辑上的表划分成多个数据块即 HRegion,并存储在 HRegionServer 中。HMaster 负责管理所有的 HRegionServer,它本身并不存储任何数据,而只是存储数据到 HRegionServer 的映射关系,即元数据。集群中的所有节点通过 ZooKeeper 进行协调,并处理 HBase 运行期间可能遇到的各种问题。HBase 架构如图 5-1 所示。

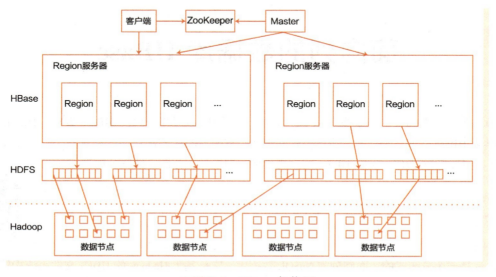

● 图 5-1　HBase 架构图

5.2.1　HBase 的组件

HBase 是一个分布式、可扩展的数据库,它提供了高效的数据存储和管理功能。通过了解 HBase 的组件,我们可以更好地理解和使用 HBase,并确保正确地配置和操作它。我们将深入探讨 HBase 的组件,并通过实际操作来掌握各个组件的作用和交互方式。

HBase 的组件包括 HBase Client、HBase Server、HMaster、RegionServer 和 ZooKeeper,它们分别负责用户接口、服务端、元数据管理、数据存储和分布式协调等不同的功能。HBase 的组件及其作用如下。

1)HBase Client:HBase Client 是 HBase 的用户接口,它提供了 Java API 和 Shell 命令行工具,用户可以通过 HBase Client 向 HBase Server 提交读写请求。

2)HBase Server:HBase Server 是 HBase 的服务端,它接收 HBase Client 提交的读写请求,并将请求转发给相应的组件进行处理。

3)HMaster:HMaster 是 HBase 的主节点,它负责管理 HBase 集群的元数据信息,如表的结构、RegionServer 的负载均衡等。

4)RegionServer:RegionServer 是 HBase 的数据节点,它负责管理 HBase 表的数据,每个 RegionServer 管理多个 Region,每个 Region 对应 HBase 表中的一个数据分区。

5）ZooKeeper：ZooKeeper 是 HBase 的分布式协调服务，它负责管理 HBase 集群的状态信息，如 RegionServer 的上下线、HMaster 的选举等。

5.2.2 HBase 工作机制

1. HMaster 工作机制

1）HMaster 启动时，会读取 HDFS 中的 ROOT 表和 META 表，获取 HBase 集群中所有表的元数据信息。

2）HMaster 会选举出一个活跃的 HMaster 节点，并将其他 HMaster 节点设置为备用状态。

3）HMaster 会定期检查 HBase 集群的状态，如 RegionServer 的上下线、HMaster 的健康状态等。

4）当 HBase 集群中新增或删除表时，HMaster 会更新 ROOT 表和 META 表，并将更新后的元数据信息同步给所有 RegionServer 节点。

5）当 HBase 集群中的 RegionServer 节点发生故障或负载过高时，HMaster 会重新分配 Region，并将 RegionServer 的负载均衡信息更新到 META 表中。

总之，HMaster 负责管理 HBase 集群的元数据信息，如表的结构、RegionServer 的负载均衡等。它会维护 ROOT 表和 META 表，更新元数据信息，并将更新后的信息同步给所有 RegionServer 节点。同时，HMaster 还会负责重新分配 Region，以实现 HBase 集群的负载均衡。

2. HRegionServer 的工作机制

1）HBase 的 HRegionServer 是数据节点，它负责管理 HBase 表的数据。每个 RegionServer 管理多个 Region，每个 Region 对应 HBase 表中的一个数据分区。当 HBase 集群启动时，每个 RegionServer 节点会向 HMaster 注册自己，并请求分配 Region。HMaster 会根据 HBase 集群的负载情况和 RegionServer 的可用性，将 Region 分配给相应的 RegionServer 节点。

2）当 HBase 表中的数据发生变化时，HRegionServer 会将变化的数据写入 WAL（Write-Ahead Log）和 MemStore 中。WAL 是 HBase 的日志文件，用于记录 HBase 表中的写操作，以保证数据的一致性和可靠性。MemStore 是 HBase 的内存缓存，用于存储 HBase 表中的写操作，以提高读写性能。当 MemStore 中的数据达到一定大小时，HRegionServer 会将数据刷写到 HDFS 中，以释放内存空间。

总之，HRegionServer 负责管理 HBase 表的数据，每个 RegionServer 管理多个 Region，每个 Region 对应 HBase 表中的一个数据分区。它会将变化的数据写入 WAL 和 MemStore 中，以保证数据的一致性和可靠性，并提高读写性能。同时，HRegionServer 会将数据刷写到 HDFS 中，以释放内存空间。

5.3 HBase 数据模型

我们将深入探讨 HBase 的数据模型，通过 HBase 的数据结构和组织方式，掌握如何使用它进行大规模数据的存储和访问。同时，你还可以学习如何优化 HBase 的性能和扩展性，以实现更高效的数据存储和管理。

HBase 是一个稀疏的、长期存储的、多维度的、排序的映射表，存在 HDFS 上。这张表的索引是行关键字、列关键字、单元格和时间戳。HBase 的数据都是字符串，没有类型。HBase 的数据模型如图 5-2 所示。

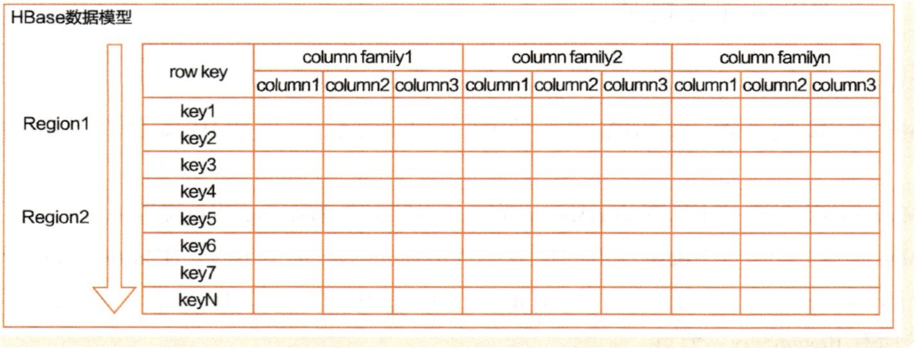

• 图 5-2 HBase 数据模型

HBase 的数据组及它们的作用如下。

1）表格（Table）：HBase 中的数据存储在表格中，表格由行和列族组成。行是表格中的基本单元，每一行都有一个唯一的行键（Row Key），列族是一组相关的列，每个列族都有一个唯一的名称。

2）列族（Column Family）：列族是表格中的一组相关的列，每个列族都有一个唯一的名称。在 HBase 中，列族是动态的，可以在表格创建后添加或删除列族。

3）列（Column）：列是表格中的一个单元格，由行键、列族和列限定符（Column Qualifier）组成。在 HBase 中，列是动态的，可以在表格创建后添加或删除列。

4）行键（Row Key）：行键是表格中的一个唯一标识符，用于标识表格中的每一行。在 HBase 中，行键是按字典序排序的，可以通过行键范围进行快速检索。

5）列限定符（Column Qualifier）：列限定符是表格中的一个标识符，用于标识表格中的每个列。在 HBase 中，列限定符是按字典序排序的，可以通过列限定符范围进行快速检索。

6）单元格版本（Cell Version）：单元格版本是表格中的一个单元格的历史记录，每个单元格可以有多个版本。在 HBase 中，单元格版本是按时间戳排序的，可以通过时间戳范围进行快速检索。

5.4 HBase 读写流程

5.4.1 HBase 写操作流程

为了快速写入数据并在海量数据中准确定位所需的数据，HBase 会将数据的元数据存储到 HMaster，然后分配具体的 HRegionServer 来存储数据，HBase 写数据的流程如图 5-3 所示。

探讨 HBase 写数据的流程，有助于了解 HBase 的数据模型和存储原理，掌握如何将数据写入

HBase 数据库，并为构建高效的数据存储系统提供重要的帮助。

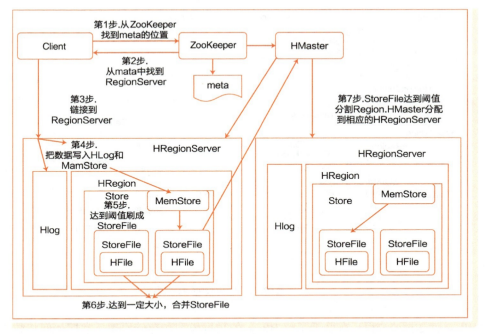

● 图 5-3　HBase 写数据的流程

1）Client 先访问 ZooKeeper，从 meta 表获取相应 Region 信息，然后找到 meta 表的数据。

2）根据 namespace、表名和 Rowkey，以及 meta 表的数据找到写入数据对应的 Region 信息。

3）找到对应的 HRegionServer。

4）把数据分别写到 HLog 和 MemStore 上一份。

5）MemStore 达到一个阈值后，则把数据刷成一个 StoreFile 文件（若 MemStore 中的数据有丢失，则可以从 HLog 上恢复）。

6）当多个 StoreFile 文件达到一定的大小后，会触发 Compact 合并操作，合并为一个 StoreFile（注意：这里同时进行版本的合并和数据删除）。

7）当 StoreFile 大小超过一定阈值后，会把当前的 Region 分割为两个 Region，并由 Hmaster 分配到相应的 HRegionServer，实现负载均衡。

5.4.2　HBase 读操作流程

HBase 为了快速读取数据，在元数据中有 RegionServer 存储的 Rowkey 范围，可以快速查找到 RegionSever，准确地说是定位到具体的 Region；其次，Region 由多个 Store 构成，Store 由多个 HFile 构成，数据存储以 block 为单位。第一步，根据每个 HFile 的 Rowkey 和 timestamp 以及布隆过滤器过滤掉不需要的 HFile，再根据每个 HFile 维护的 block 索引结合 Rowkey 定位到 block 块。HBase 读取数据的流程如图 5-4 所示。

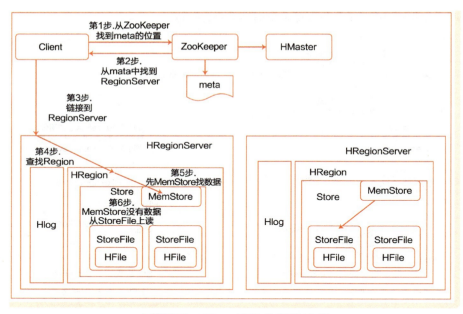

• 图 5-4　HBase 读取数据的流程

探讨 HBase 读数据的流程，有助于了解 HBase 的数据模型和存储原理，掌握如何将数据读取入 HBase 数据库，并为构建高效的读取数据存系统提供重要的帮助。

1）首先客户端从 ZooKeeper 找到 meta 表的 Region 的位置，然后读取 meta 表中的数据。而 meta 中又存储了用户表的 Region 信息。

2）根据 meta 表中的信息找到写入数据的 Namespace、表名和 Rowkey 对应的 Region 信息。

3）客户端找到对应的 HRegionServer。

4）在 HRegionServer 内查找对应的 Region。

5）从 MemStore 找数据。

6）如果 MemStore 没有数据，再到 StoreFile 上读，然后将合并后的最终结果返回给客户端。

5.5　HBase 的 API 示例

在这个教程中，我们将通过实际示例来展示如何使用 HBase 的 API 进行数据的读写操作。通过本实战，你将了解 HBase 的 API 种类和用法，并掌握如何将它们应用于实际业务。

▶▶ 5.5.1　HBase 创建表

使用 HBase 创建表操作的示例代码。

```
import org.apache.hadoop.conf.Configuration;
import org.apache.hadoop.hbase.HBaseConfiguration;
```

```java
import org.apache.hadoop.hbase.TableName;
import org.apache.hadoop.hbase.client.Admin;
import org.apache.hadoop.hbase.client.Connection;
import org.apache.hadoop.hbase.client.ConnectionFactory;
import org.apache.hadoop.hbase.client.TableDescriptor;
import org.apache.hadoop.hbase.client.TableDescriptorBuilder;
import org.apache.hadoop.hbase.util.Bytes;

import java.io.IOException;

public class HBaseCreateTable {
    public static void main(String[] args) throws IOException {
        // 创建一个配置对象
        Configuration conf = HBaseConfiguration.create();
        // 设置 HBase 集群的 Zookeeper 地址
        conf.set("hbase.zookeeper.quorum", "localhost");
        // 创建一个连接对象
        Connection conn = ConnectionFactory.createConnection(conf);
        // 获取一个 Admin 对象
        Admin admin = conn.getAdmin();
        // 创建一个表描述符对象
        TableDescriptor tableDescriptor = TableDescriptorBuilder.newBuilder(TableName.valueOf("test_table"))
                .setColumnFamily(ColumnFamilyDescriptorBuilder.of(Bytes.toBytes("cf")))
                .build();
        // 创建表
        admin.createTable(tableDescriptor);
        // 关闭 Admin 对象和连接对象
        admin.close();
        conn.close();
    }
}
```

在这段代码中，我们首先创建了一个 Configuration 对象，并设置了 HBase 集群的 ZooKeeper 地址。然后，使用 ConnectionFactory 类的 createConnection 方法创建了一个连接对象。接着，获取了一个 Admin 对象，用于创建表。我们使用 TableDescriptorBuilder 类的 newBuilder 方法创建了一个表描述符对象，并设置了表名和列族。最后，使用 Admin 对象的 createTable 方法创建表。

▶▶ 5.5.2 HBase 保持数据

使用 HBase 保持数据的示例代码。

```java
import org.apache.hadoop.conf.Configuration;
import org.apache.hadoop.hbase.HBaseConfiguration;
import org.apache.hadoop.hbase.TableName;
import org.apache.hadoop.hbase.client.*;
```

```
import org.apache.hadoop.hbase.util.Bytes;

import java.io.IOException;

public class HBasePutData {
    public static void main(String[] args) throws IOException {
        // 创建一个配置对象
        Configuration conf = HBaseConfiguration.create();
        // 设置 HBase 集群的 Zookeeper 地址
        conf.set("hbase.zookeeper.quorum", "localhost");
        // 创建一个连接对象
        Connection conn = ConnectionFactory.createConnection(conf);
        // 获取一个 Table 对象
        Table table = conn.getTable(TableName.valueOf("test_table"));
        // 创建一个 Put 对象
        Put put = new Put(Bytes.toBytes("row1"));
        // 向 Put 对象中添加数据
        put.addColumn(Bytes.toBytes("cf"), Bytes.toBytes("col1"), Bytes.toBytes("value1"));
        // 将数据存入 HBase 表中
        table.put(put);
        // 关闭 Table 对象和连接对象
        table.close();
        conn.close();
    }
}
```

在这段代码中，我们首先创建了一个 Configuration 对象，并设置了 HBase 集群的 ZooKeeper 地址。然后，使用 ConnectionFactory 类的 createConnection 方法创建了一个连接对象。接着，获取了一个 Table 对象，用于操作表中的数据。我们使用 Put 类的构造方法创建了一个 Put 对象，并向其中添加了一行数据。最后，使用 Table 对象的 put 方法将数据存入 HBase 表中。

5.5.3 HBase 更新数据

使用 HBase 更新数据的代码。

```
import org.apache.hadoop.conf.Configuration;
import org.apache.hadoop.hbase.HBaseConfiguration;
import org.apache.hadoop.hbase.TableName;
import org.apache.hadoop.hbase.client.*;
import org.apache.hadoop.hbase.util.Bytes;

import java.io.IOException;

public class HBasePutData {
    public static void main(String[] args) throws IOException {
        // 创建一个配置对象
```

```
        Configuration conf = HBaseConfiguration.create();
        // 设置 HBase 集群的 Zookeeper 地址
        conf.set("hbase.zookeeper.quorum", "localhost");
        // 创建一个连接对象
        Connection conn = ConnectionFactory.createConnection(conf);
        // 获取一个 Table 对象
        Table table = conn.getTable(TableName.valueOf("test_table"));
        // 创建一个 Put 对象
        Put put = new Put(Bytes.toBytes("row1"));
        // 向 Put 对象中添加数据
        put.addColumn(Bytes.toBytes("cf"), Bytes.toBytes("col1"), Bytes.toBytes("value1"));
        // 将数据存入 HBase 表中
        table.put(put);
        // 关闭 Table 对象和连接对象
        table.close();
        conn.close();
    }
}
```

在这段代码中,我们首先创建了一个 Configuration 对象,并设置了 HBase 集群的 ZooKeeper 地址。然后,使用 ConnectionFactory 类的 createConnection 方法创建了一个连接对象。接着,获取了一个 Table 对象,用于操作表中的数据。我们使用 Put 类的构造方法创建了一个 Put 对象,并向其中添加了一行数据。然后,使用 Put 对象的 addColumn 方法更新了该行数据的某个列的值。最后,使用 Table 对象的 put 方法将数据更新到 HBase 表中。

5.5.4 HBase 获取数据

使用 HBase 获取数据的代码。

```
import org.apache.hadoop.conf.Configuration;
import org.apache.hadoop.hbase.HBaseConfiguration;
import org.apache.hadoop.hbase.TableName;
import org.apache.hadoop.hbase.client.*;
import org.apache.hadoop.hbase.util.Bytes;

import java.io.IOException;

public class HBaseGetData {
    public static void main(String[] args) throws IOException {
        // 创建一个配置对象
        Configuration conf = HBaseConfiguration.create();
        // 设置 HBase 集群的 Zookeeper 地址
        conf.set("hbase.zookeeper.quorum", "localhost");
        // 创建一个连接对象
        Connection conn = ConnectionFactory.createConnection(conf);
        // 获取一个 Table 对象
```

```java
        Table table = conn.getTable(TableName.valueOf("test_table"));
        // 创建一个 Get 对象
        Get get = new Get(Bytes.toBytes("row1"));
        // 从 HBase 表中获取数据
        Result result = table.get(get);
        // 遍历结果集
        for (Cell cell : result.listCells()) {
            // 获取行键
            String rowKey = Bytes.toString(CellUtil.cloneRow(cell));
            // 获取列族
            String family = Bytes.toString(CellUtil.cloneFamily(cell));
            // 获取列名
            String qualifier = Bytes.toString(CellUtil.cloneQualifier(cell));
            // 获取值
            String value = Bytes.toString(CellUtil.cloneValue(cell));
            // 输出结果
            System.out.printf("rowKey=%s, family=%s, qualifier=%s, value=%s%n", rowKey, family, qualifier, value);
        }
        // 关闭 Table 对象和连接对象
        table.close();
        conn.close();
    }
}
```

在这段代码中，我们首先创建了一个 Configuration 对象，并设置了 HBase 集群的 ZooKeeper 地址。然后，使用 ConnectionFactory 类的 createConnection 方法创建了一个连接对象。接着，获取了一个 Table 对象，用于操作表中的数据。我们使用 Get 类的构造方法创建了一个 Get 对象，并指定了要获取的行键。然后，使用 Table 对象的 get 方法从 HBase 表中获取数据，并将结果存储在 Result 对象中。最后，遍历 Result 对象中的 Cell 对象，获取行键、列族、列名和值，并输出结果。

▶▶ 5.5.5　HBase 删除数据

使用 HBase 删除数据的代码。

```java
import org.apache.hadoop.conf.Configuration;
import org.apache.hadoop.hbase.HBaseConfiguration;
import org.apache.hadoop.hbase.TableName;
import org.apache.hadoop.hbase.client.*;
import org.apache.hadoop.hbase.util.Bytes;

import java.io.IOException;

public class HBaseDeleteData {
    public static void main(String[] args) throws IOException {
        // 创建一个配置对象
```

```
        Configuration conf = HBaseConfiguration.create();
        // 设置 HBase 集群的 Zookeeper 地址
        conf.set("hbase.zookeeper.quorum", "localhost");
        // 创建一个连接对象
        Connection conn = ConnectionFactory.createConnection(conf);
        // 获取一个 Table 对象
        Table table = conn.getTable(TableName.valueOf("test_table"));
        // 创建一个 Delete 对象
        Delete delete = new Delete(Bytes.toBytes("row1"));
        // 从 HBase 表中删除数据
        table.delete(delete);
        // 关闭 Table 对象和连接对象
        table.close();
        conn.close();
    }
}
```

在这段代码中，我们首先创建了一个 Configuration 对象，并设置了 HBase 集群的 ZooKeeper 地址。然后，使用 ConnectionFactory 类的 createConnection 方法创建了一个连接对象。接着，获取了一个 Table 对象，用于操作表中的数据。我们使用 Delete 类的构造方法创建了一个 Delete 对象，并指定了要删除的行键。然后，使用 Table 对象的 delete 方法从 HBase 表中删除数据。最后，关闭了 Table 对象和连接对象。

▶▶ 5.5.6 使用 HBase 获取某一行数据

使用 HBase 获取某一行数据的代码。

```
import org.apache.hadoop.hbase.HBaseConfiguration;
import org.apache.hadoop.hbase.TableName;
import org.apache.hadoop.hbase.client.*;
import org.apache.hadoop.hbase.util.Bytes;

import java.io.IOException;

public class HBaseGetRowData {
    public static void main(String[] args) throws IOException {
        // 创建一个配置对象
        Configuration conf = HBaseConfiguration.create();
        // 设置 HBase 集群的 Zookeeper 地址
        conf.set("hbase.zookeeper.quorum", "localhost");
        // 创建一个连接对象
        Connection conn = ConnectionFactory.createConnection(conf);
        // 获取一个 Table 对象
        Table table = conn.getTable(TableName.valueOf("test_table"));
        // 创建一个 Get 对象
        Get get = new Get(Bytes.toBytes("row1"));
```

```
        // 获取一行数据
        Result result = table.get(get);
        // 输出行键和列族、列名、列值
        byte[] row = result.getRow();
        byte[] value = result.getValue(Bytes.toBytes("cf"), Bytes.toBytes("col1"));
        System.out.println("Row: " + Bytes.toString(row) + " Value: " + Bytes.toString(value));
        // 关闭 Table 对象和连接对象
        table.close();
        conn.close();
    }
}
```

在这段代码中,我们首先创建了一个 Configuration 对象,并设置了 HBase 集群的 ZooKeeper 地址。然后,使用 ConnectionFactory 类的 createConnection 方法创建了一个连接对象。接着,获取了一个 Table 对象,用于操作表中的数据。我们创建了一个 Get 对象,并设置了要获取的行键。然后,使用 Table 对象的 get 方法获取一行数据,并将结果存储在一个 Result 对象中。最后,输出了行键和列族、列名、列值。

5.6 HBase 存储订单案例

使用 Spring Boot 集成 HBase 的项目。在这个项目中,我们将创建一个订单存储系统,其中包含了订单的基本信息,例如订单号、订单状态、订单金额等。按照以下步骤进行操作。

1) 在 pom.xml 文件中添加依赖项。

```xml
<?xml version="1.0" encoding="UTF-8"? >
<project xmlns="http://maven.apache.org/POM/4.0.0"
    xmlns:xsi="http://www.w3.org/2001/XMLSchema-instance"
    xsi:schemaLocation="http://maven.apache.org/POM/4.0.0 http://maven.apache.org/xsd/maven-4.0.0.xsd">
    <modelVersion>4.0.0</modelVersion>

    <groupId>com.it.jareny.bigdata</groupId>
    <artifactId>jareny-bigdata-hbase</artifactId>
    <version>1.0-SNAPSHOT</version>

    <parent>
        <groupId>org.springframework.boot</groupId>
        <artifactId>spring-boot-starter-parent</artifactId>
        <version>2.2.5.RELEASE</version>
    </parent>

    <dependencies>
        <!-- SpringBoot 相关依赖 -->
        <dependency>
```

```xml
    <groupId>org.springframework.boot</groupId>
    <artifactId>spring-boot-starter</artifactId>
</dependency>
<dependency>
    <groupId>org.springframework.boot</groupId>
    <artifactId>spring-boot-starter-web</artifactId>
</dependency>
<!--HBase 相关依赖 -->
<dependency>
    <groupId>org.apache.hbase</groupId>
    <artifactId>hbase-server</artifactId>
    <version>1.3.1</version>
</dependency>
<dependency>
    <groupId>org.apache.hbase</groupId>
    <artifactId>hbase-client</artifactId>
    <version>1.3.1</version>
</dependency>
<dependency>
    <groupId>org.springframework.data</groupId>
    <artifactId>spring-data-hadoop-hbase</artifactId>
    <version>2.5.0.RELEASE</version>
</dependency>
<dependency>
    <groupId>org.springframework.data</groupId>
    <artifactId>spring-data-hadoop</artifactId>
    <version>2.5.0.RELEASE</version>
</dependency>
<dependency>
    <groupId>org.apache.hadoop</groupId>
    <artifactId>hadoop-hdfs</artifactId>
    <version>2.7.3</version>
</dependency>
<dependency>
    <groupId>org.springframework.data</groupId>
    <artifactId>spring-data-hadoop-core</artifactId>
    <version>2.4.0.RELEASE</version>
</dependency>
<dependency>
    <groupId>org.apache.hbase</groupId>
    <artifactId>hbase</artifactId>
    <version>1.3.2</version>
    <type>pom</type>
</dependency>
<dependency>
    <groupId>org.projectlombok</groupId>
```

```xml
            <artifactId>lombok</artifactId>
            <version>1.18.16</version>
        </dependency>
</project>
```

2) 在 application.properties 文件中添加 HBase 配置。

```
server.port=60003
spring.application.name=jareny-bigdata-server
spring.data.hbase.quorum=192.168.81.111:2181
spring.data.hbase.rootDir=hdfs://192.168.81.111:9000/hbase
```

3) 创建一个 HBaseConfiguration 类，用于配置 HBaseTemplate。

```java
package com.jareny.bigdata.hbase.config;

import org.apache.hadoop.hbase.HBaseConfiguration;
import org.springframework.beans.factory.annotation.Value;
import org.springframework.context.annotation.Bean;
import org.springframework.context.annotation.Configuration;
import org.springframework.data.hadoop.hbase.HbaseTemplate;

@Configuration
public class HBaseConfig {
    @Value("${spring.data.hbase.quorum}")
    private String quorum;

    @Value("${spring.data.hbase.rootDir}")
    private String rootDir;

    @Bean
    public HbaseTemplate hbaseTemplate() {
        org.apache.hadoop.conf.Configuration configuration = HBaseConfiguration.create();
        configuration.set("hbase.zookeeper.quorum", quorum);
        configuration.set("hbase.rootdir", rootDir);
        return new HbaseTemplate(configuration);
    }
}
```

4) 创建一个 HBaseRepository 类，用于访问 HBase 表。

```java
package com.jareny.bigdata.hbase.dao;

import com.jareny.bigdata.hbase.entity.Order;
import org.apache.hadoop.hbase.client.Get;
import org.apache.hadoop.hbase.client.Put;
import org.apache.hadoop.hbase.client.Result;
import org.apache.hadoop.hbase.util.Bytes;
import org.springframework.beans.factory.annotation.Autowired;
```

```java
import org.springframework.data.hadoop.hbase.HbaseTemplate;
import org.springframework.stereotype.Repository;
import java.io.IOException;

@Repository
public class HBaseRepository {

    @Autowired
    private HbaseTemplate hbaseTemplate;
    public void put(String tableName, String rowKey, String columnFamily, String column, String value) {
        hbaseTemplate.execute(tableName, table -> {
            Put put = new Put(Bytes.toBytes(rowKey));
            put.addColumn(Bytes.toBytes(columnFamily), Bytes.toBytes(column), Bytes.toBytes(value));
            table.put(put);
            return null;
        });
    }

    public String get(String tableName, String rowKey, String columnFamily, String column) {
        return hbaseTemplate.get(tableName, rowKey, columnFamily, column, (result, rowNum) -> {
            byte[] valueBytes = result.getValue(Bytes.toBytes(columnFamily), Bytes.toBytes(column));
            return Bytes.toString(valueBytes);
        });
    }

    public void saveOrder(String tableName, Order order) throws IOException {
        Put put = order.toPut();
        hbaseTemplate.execute(tableName, table -> {
            table.put(put);
            return null;
        });
    }

    public Order getOrderById(String tableName, String columnFamily, String orderId) throws IOException {
        Get get = new Get(Bytes.toBytes(orderId));
        Result result = hbaseTemplate.execute(tableName, table -> table.get(get));
        if (result.isEmpty()) {
            return null;
        }
        String customer = Bytes.toString(result.getValue(Bytes.toBytes(columnFamily), Bytes.toBytes("customer")));
        String product = Bytes.toString(result.getValue(Bytes.toBytes(columnFamily), Bytes.toBytes("product")));
```

```java
        String quantity = Bytes.toString(result.getValue(Bytes.toBytes(columnFamily),
Bytes.toBytes("quantity")));
        String amount = Bytes.toString(result.getValue(Bytes.toBytes(columnFamily), Bytes.
toBytes("amount")));
        String status = Bytes.toString(result.getValue(Bytes.toBytes(columnFamily), Bytes.
toBytes("status")));
        return new Order(orderId, customer, product, quantity, amount, status);
    }
}
```

5) 创建一个 Controller 类，使用它来存储和获取订单。

```java
package com.jareny.bigdata.hbase.controller;

import com.jareny.bigdata.hbase.dao.HBaseRepository;
import com.jareny.bigdata.hbase.entity.Order;
import org.springframework.beans.factory.annotation.Autowired;
import org.springframework.web.bind.annotation.*;

@RestController
public class OrderController {
    @Autowired
    private HBaseRepository hbaseRepository;

    @PostMapping("/orders/create")
    public void create(@RequestBody Order order) {
        hbaseRepository.put("orders", order.getOrderId(), "details", "customer", order.getCustomer());
        hbaseRepository.put("orders", order.getOrderId(), "details", "product", order.getProduct());
        hbaseRepository.put("orders", order.getOrderId(), "details", "quantity", order.getQuantity());
        hbaseRepository.put("orders", order.getOrderId(), "details", "amount", order.getAmount());
        hbaseRepository.put("orders", order.getOrderId(), "details", "status", order.getStatus());
    }

    @GetMapping("/orders/{id}")
    public Order get(@PathVariable String id) {
        String customer = hbaseRepository.get("orders", id, "details", "customer");
        String product = hbaseRepository.get("orders", id, "details", "product");
        String quantity = hbaseRepository.get("orders", id, "details", "quantity");
        String amount = hbaseRepository.get("orders", id, "details", "amount");
        String status = hbaseRepository.get("orders", id, "details", "status");
        return new Order(id, customer, product, quantity, amount, status);
    }
```

```
    @PostMapping("/orders/save")
    public void save(@RequestBody Order order) throws Exception {
        hbaseRepository.saveOrder("orders", order);
    }

    @GetMapping("/orders/{id}")
    public void getOrder(@PathVariable String id) throws Exception {
        hbaseRepository.getOrderById("orders", "details", id);
    }
}
```

6）创建一个 HBaseApplication 类，使用它启动项目。

```
package com.jareny.bigdata.hbase;

import org.springframework.boot.SpringApplication;
import org.springframework.boot.autoconfigure.SpringBootApplication;

@SpringBootApplication
public class HBaseApplication {
    public static void main(String[] args) {
        SpringApplication.run(HBaseApplication.class,args);
    }
}
```

第6章

数据迁移工具 Sqoop

本章内容介绍 Sqoop 的基本概念和工作原理,包括数据传输、导入和导出等步骤。学习 Sqoop 的配置文件格式和各参数的作用,包括连接数据库、选择表、指定导入导出格式等。掌握 Sqoop 的常用命令和操作,包括导入、导出、增量导入等操作。学习 Sqoop 的高级特性和应用场景,包括数据过滤、转换、分区等功能。实践 Sqoop 的应用案例,包括数据传输等场景。

6.1 Sqoop 架构和工作原理

Sqoop 是一个用于在 Hadoop 和结构化数据存储(如关系型数据库)之间进行大规模数据迁移的工具。通过了解 Sqoop 的架构和工作原理,我们可以更好地理解和使用 Sqoop,并确保正确地设计和操作它。Sqoop 是一种用于在 Hadoop 和关系型数据库之间进行数据传输的工具,Sqoop 的架构如图 6-1 所示。

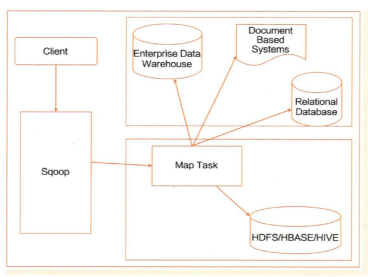

● 图 6-1 Sqoop 架构图

Sqoop 的架构概述，包括其组成部分和各部分的功能。Sqoop 的架构包括连接器、导入器、连接器元数据存储以及命令行接口等部分。了解这些组成部分将帮助你更好地理解 Sqoop 的整体架构和工作原理。Sqoop 的组件以及作用如下。

1）Sqoop Client：Sqoop Client 是 Sqoop 的客户端，它提供了命令行界面和 API 接口，用于与 Sqoop Server 进行交互。Sqoop Client 可以在本地或远程运行，可以进行数据的导入（import）和导出（export）。

2）Sqoop Server：Sqoop Server 是 Sqoop 的服务器，它负责接收 Sqoop Client 的请求，并将请求转发给相应的 Sqoop Connector 进行处理。Sqoop Server 可以在本地或远程运行，可以进行多用户的并发处理。

3）Sqoop Connector：Sqoop Connector 是 Sqoop 的连接器，它负责连接 Hadoop 和关系型数据库之间的数据传输。Sqoop Connector 可以根据不同的数据库类型和版本进行配置，也可以进行数据的分片和并行传输。

4）Hadoop：Hadoop 是一个开源的分布式计算框架，它可以处理大规模数据的存储和计算。Sqoop 可以将关系型数据库中的数据导入到 Hadoop 中进行处理，也可以将 Hadoop 中的数据导出到关系型数据库中进行存储。

5）关系型数据库：关系型数据库是一种基于关系模型的数据库，它使用表（table）来存储数据。Sqoop 可以将关系型数据库中的数据导入到 Hadoop 中进行处理，也可以将 Hadoop 中的数据导出到关系型数据库中进行存储。

Sqoop 的工程流程：用户向 Sqoop 发起一个命令之后，这个命令会转换为一个基于 Map Task 的 MapReduce 作业。Map Task 会访问数据库的元数据信息，通过并行的 Map Task 将数据库的数据读取出来，然后导入 Hadoop 中。当然也可以将 Hadoop 中的数据导入传统的关系型数据库中。它的核心思想就是通过基于 Map Task 的 MapReduce 作业，实现数据的并发拷贝和传输，这样可以大大提高效率。

▶▶ 6.1.1 Sqoop 导入原理

Sqoop 的导入原理是将关系型数据库中的数据读取到内存中，进行分片和并行传输，将数据传输到 Hadoop 的分布式文件系统（HDFS）中，然后进行数据的处理和计算，最后将处理结果写入 HDFS 中，如图 6-2 所示。

Sqoop 导入运行流程如下。

1）Sqoop Client 向 Sqoop Server 发送导入请求。

2）Sqoop Server 接收到导入请求后，将请求转发给相应的 Sqoop Connector 进行处理。

3）Sqoop Connector 连接关系型数据库，并执行相应的 SQL 语句，将数据读取到内存中。

4）Sqoop Connector 将读取到的数据进行分片和并行传输，将数据传输到 Hadoop 的分布式文件系统（HDFS）中。

5）Sqoop 将 HDFS 中的数据进行处理，生成相应的 MapReduce 任务，进行数据的处理和计算。

6）处理完成后，Sqoop 将处理结果写入 HDFS 中。

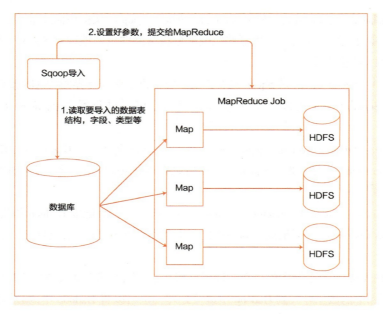

• 图 6-2 Sqoop 导入原理

6.1.2 Sqoop 导出原理

Sqoop 的导出原理是将 Hadoop 分布式文件系统中的数据读取到内存中，进行分片和并行传输，然后进行数据的处理和计算，最后将处理结果写入关系型数据库中，如图 6-3 所示。

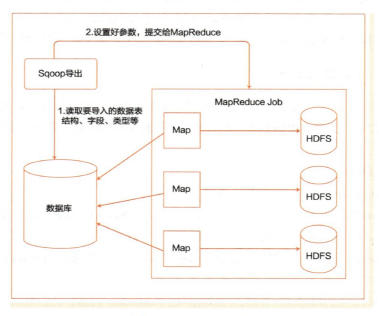

• 图 6-3 Sqoop 导出原理

Sqoop 导出运行流程如下。

1）Sqoop Client 向 Sqoop Server 发送导出请求。

2）Sqoop Server 接收到导出请求后，将请求转发给相应的 Sqoop Connector 进行处理。

3）Sqoop Connector 连接关系型数据库，并执行相应的 SQL 语句，将数据读取到内存中。

4）Sqoop Connector 将读取到的数据进行分片和并行传输，将数据传输到 Hadoop 的分布式文件系统（HDFS）中。

5）Sqoop 将 HDFS 中的数据进行处理，生成相应的 MapReduce 任务，进行数据的处理和计算。

6）处理完成后，Sqoop 将处理结果写入关系型数据库中。

6.2 Sqoop 将 HDFS 数据导入 MySQL

在以下的实战操作中，我们将深入探讨 Sqoop 将 HDFS 数据导入 MySQL 的流程，并通过实际操作来掌握如何使用 Sqoop 实现数据迁移。

Sqoop 将 HDFS 数据导入 MySQL 的基本步骤：创建 MySQL 表、生成 Sqoop 命令、执行 Sqoop 命令、验证导入的数据。

1）在 MySQL 数据库中新建表，如下。

```
CREATE TABLE UserOrderInfo(
orderId     bigint          COMMENT '订单id',
userId      bigint          COMMENT '用户id',
productId   bigint          COMMENT '产品id',
orderTime   varchar(100)    COMMENT '下单时间',
payAmount   double          COMMENT '付款金额'
);
```

2）在 HDFS 上创建目录，如下。

```
[root@Linux-1 sbin]# hadoop fs -mkdir -p /sqoop/export/
```

3）执行模拟数据的程序，如下。

```
package com.jareny.bigdata.mock.model;

import lombok.AllArgsConstructor;
import lombok.Data;
import lombok.NoArgsConstructor;
import java.io.File;
import java.io.FileNotFoundException;
import java.io.PrintWriter;
import java.time.LocalDateTime;
import java.time.ZoneOffset;
import java.time.format.DateTimeFormatter;
import java.util.ArrayList;
```

```java
import java.util.List;
import java.util.Random;

// 第 9 章的实战数据模拟
@Data
@AllArgsConstructor
@NoArgsConstructor
@SuppressWarnings("all")
public class UserOrderInfo {
    private Integer orderId      ; //  '订单 id',
    private Integer userId       ; //  '用户 id',
    private Integer productId    ; //  '产品 id',
    private Long orderTime       ; //  '下单时间',
    private double payAmount     ; //  '付款金额'

    @Override
    public String toString() {
        StringBuffer order = new StringBuffer();
        order.append(orderId).append("\t");
        order.append(userId).append("\t");
        order.append(productId).append("\t");
        order.append(payAmount).append("\t");
        order.append(orderTime).append("\t");
        return order.toString();
    }

    // 假如有 10000 人
    private static final int  userIds = 10000;
    // 假如有 100000 商品
    private static final int  productIds = 100000;
    // 模拟 200,0000 条数据
    private static final int  rowMax = 12000;

    // 模拟数据
    public static List<UserOrderInfo> getRecordList(){
        List<UserOrderInfo> recordList = new ArrayList<>();
        // 基准时间
        LocalDateTime localDateTime = LocalDateTime.now();
        DateTimeFormatter formatter = DateTimeFormatter.ofPattern("yyyy-MM-dd HH:mm:ss");
        Random random = new Random();
        for (int rowIndex = 10000;rowIndex<rowMax;rowIndex++){
            UserOrderInfo order = new UserOrderInfo();
            order.setUserId(Integer.valueOf((int) (Math.random() * random.nextInt(userIds))));
            order.setOrderId(rowIndex);
            order.setProductId(random.nextInt(productIds)+random.nextInt(productIds));
            long orderTime = LocalDateTime.now()
```

```
                .minusYears(random.nextInt(5))
                .minusMonths(random.nextInt(12))
                .minusDays(random.nextInt(30))
                .minusHours(random.nextInt(24))
                .minusMinutes(random.nextInt(60))
                .minusSeconds(random.nextInt(60))
                .toInstant(ZoneOffset.of("+8")).toEpochMilli();
        order.setOrderTime(orderTime);
        order.setPayAmount(random.nextInt(1000)+random.nextInt(500));
        recordList.add(order);
    }
    return recordList;
}

// 写入文件
public static <T> void writeToFile(List<T> list) throws FileNotFoundException {
    String path = System.getProperty("user.dir");
    File file = new File("D:\\jareny\\MockData\\mapreduce\\UserOrderInfo.txt");
    PrintWriter printWriter = new PrintWriter(file);
    list.stream().forEach(value->printWriter.println(value.toString()));
    printWriter.flush();
    printWriter.close();
}

// 启动模拟数据
public static void main(String[] args) throws FileNotFoundException {
    long start = System.currentTimeMillis();
    List<UserOrderInfo> recordList = getRecordList();
    writeToFile(recordList);
    long end = System.currentTimeMillis();
    System.out.println("模拟数据完成!,耗时:"+(end-start)+"毫秒");
    System.out.println(System.getProperty("user.dir"));
}
```

4）将模拟数据文件上传到 HDFS。

将文件通过工具上传到服务器的 **/opt/jareny/bigdata/data/UserOrderInfo.txt**，然后使用 Hadoop 的命令上传到 HDFS 上，如下。

```
[root@Linux-1 sbin]# hadoop fs -copyFromLocal /opt/jareny/bigdata/data/UserOrderInfo.txt /sqoop/export/
```

5）执行如下命令导出。

```
[root@Linux-1 sqoop]# bin/sqoop export \
--connect jdbc:mysql://Linux-1:3306/jarenyBigdata \
--username root \
```

```
--password 123456 \
--table  UserOrderInfo \
--export-dir /sqoop/export/   \
--fields-terminated-by '\t'
```

6）查看导出到 MySQL 的数据。

```
mysql> SELECT * FROM UserOrderInfo limit 5;
+---------+--------+----------+----------+---------------+
|orderId |userId |productId |orderTime |payAmount      |
+---------+--------+----------+----------+---------------+
| 10000  | 4368  |   126624 |941.0     |1507940692590 |
| 10001  | 1593  |    97616 |706.0     |1486285591590 |
| 10002  |   75  |   134528 |262.0     |1492177408591 |
| 10003  | 5095  |    96462 |778.0     |1569404722591 |
| 10004  |  150  |    63082 |219.0     |1526191638591 |
+---------+--------+----------+----------+---------------+
```

6.3 Sqoop 将 MySQL 数据导入 HDFS

在以下的实战操作中，我们将深入探讨 Sqoop 将 MySQL 数据导入 HDFS 的流程，并通过实际操作来掌握如何使用 Sqoop 实现数据迁移。

Sqoop 将 MySQL 数据导入 HDFS 的基本步骤：创建 HDFS 目录、生成 Sqoop 命令、执行 Sqoop 命令验证导入的数据。

1. 普通导入

1）首先看看 MySQL 的表格。

```
mysql>  SELECT * FROM UserOrderInfo limit 5;
```

2）在 Sqoop 安装的目录下，执行命令。

```
[root@Linux-1 sqoop]# bin/sqoop import \
--connect jdbc:mysql://Linux-1:3306/jarenyBigdata \
--username root \
--password 123456 \
--table UserOrderInfo \
--m 1
```

3）查看导入的数据。

```
[root@Linux-1 sqoop]# hadoop fs -cat /user/root/UserOrderInfo/part-m-00000
11024,1750,82133,1235.0,1.476253946604E12
11025,317,85708,1021.0,1.511497991604E12
11026,1435,126784,533.0,1.616235187604E12
11027,1838,102535,470.0,1.556911882604E12
11028,6164,169045,551.0,1.463022529604E12
```

```
11029,5302,63950,162.0,1.458769430604E12
11030,3451,137123,437.0,1.517022131604E12
11031,339,70926,157.0,1.469321992604E12
11032,5371,89584,1105.0,1.579594273604E12
11033,2451,140679,718.0,1.480569443604E12
11034,9600,106874,1146.0,1.539091922604E12
```

2. 指定导入路径和分隔符

1) 在 Sqoop 安装的目录下，执行如下命令。

```
[root@Linux-1 sqoop]# bin/sqoop import \
--connect jdbc:mysql://Linux-1:3306/jarenyBigdata \
--username root  \
--password 123456 \
--table UserOrderInfo \
--target-dir /sqoop/import/hdfs/UserOrderInfo \
--fields-terminated-by '\t' \
-m 1
```

2) 执行后，查询执行的结果。

```
[root@Linux-1 sbin]# hadoop fs -cat /sqoop/import/hdfs/UserOrderInfo/part-m-00000
11984   4144    110336   769.0    1.471538368609E12
11985   2245    152090   1004.0   1.605498271609E12
11986   1886    123869   634.0    1.605747102609E12
11987   2112    92567    987.0    1.582728341609E12
11988   139     104054   1374.0   1.611904423609E12
11989   3418    186152   1090.0   1.515642325609E12
11990   2311    73344    840.0    1.536582041609E12
11991   688     157500   542.0    1.573900874609E12
11992   142     176352   426.0    1.579264348609E12
11993   2753    128827   752.0    1.505207926609E12
11994   262     86578    662.0    1.579439068609E12
11995   2128    45577    908.0    1.470092963609E12
11996   4261    91819    1140.0   1.492191705609E12
11997   273     132842   1095.0   1.520562290609E12
11998   2239    94304    562.0    1.569316586609E12
11999   1597    149852   548.0    1.556486774609E12
```

3. 导入 where 条件的数据

用户可以导入表的一部分数据，这通常是通过在 SQL 查询中使用 where 子句来实现的。Sqoop 在数据库服务器上执行相应的 SQL 查询，并将结果存储在 HDFS 的目标目录中。

1) 执行如下命令。

```
[root@Linux-1 sqoop] bin/sqoop import   \
--connect jdbc:mysql://Linux-1:3306/jarenyBigdata \
--username root   \
```

```
--password 123456 \
--where payAmount>=1000.0 \
--table UserOrderInfo \
--target-dir /sqoop/import/hdfs/where \
-m 1
```

2）执行完成后，查看结果。

```
[root@Linux-1 hadoop-2.7.3]# hadoop fs -cat /sqoop/import/hdfs/where/part-m-00000
11983,3033,60352,1100.0,1.540831560609E12
11984,4144,110336,769.0,1.471538368609E12
11985,2245,152090,1004.0,1.605498271609E12
11986,1886,123869,634.0,1.605747102609E12
11987,2112,92567,987.0,1.582728341609E12
11988,139,104054,1374.0,1.611904423609E12
```

4. 导入 Query 结果数据

1）执行如下命令。

```
[root@Linux-1 sqoop]# bin/sqoop import \
--connect jdbc:mysql://Linux-1:3306/jarenyBigdata \
--username root \
--password 123456 \
--target-dir /sqoop/import/hdfs/query \
--query 'SELECT * FROM UserOrderInfo WHERE payAmount>=1000.0 and $CONDITIONS' \
--fields-terminated-by '\t' \
--m 1
```

2）查询数据。

```
[root@Linux-1 hadoop-2.7.3]# hadoop dfs -cat /sqoop/import/hdfs/query/part-m-00000
11990    2311    73344     840.0     1.536582041609E12
11991    688     157500    542.0     1.573900874609E12
11992    142     176352    426.0     1.579264348609E12
11993    2753    128827    752.0     1.505207926609E12
11994    262     86578     662.0     1.579439068609E12
11995    2128    45577     908.0     1.470092963609E12
11996    4261    91819     1140.0    1.492191705609E12
11997    273     132842    1095.0    1.520562290609E12
11998    2239    94304     562.0     1.569316586609E12
11999    1597    149852    548.0     1.556486774609E12
```

第 7 章 数据采集工具 Flume

本章内容介绍 Flume 的基本概念和工作原理，包括数据采集、聚合和传输等步骤。学习 Flume 的配置文件格式和各组件的作用，包括 Source、Channel、Sink 和 Interceptor 组件。掌握 Flume 的常用命令和操作，包括启动、停止、重启等操作。学习 Flume 的高级特性和应用场景，包括数据过滤、转换、分发等功能。编写 Flume 的应用案例，包括日志采集、数据传输等场景。总之，学习 Flume 可以帮助我们更好地处理和收集日志数据，从而实现更好的数据分析和处理。

7.1 Flume 简介

1. Flume 的概述

Apache Flume 是一个可以收集日志、事件等数据资源，并将这些数量庞大的数据从各项数据资源中集中起来存储的工具。Flume 具有高可用、分布式、配置工具的特点，其设计的原理也是基于将数据流（如日志数据）从各种网站服务器上汇集起来存储到 HDFS、HBase 等集中存储器中。

2. Flume 的特性

1）可靠性：当节点出现故障时，日志能够被传送到其他节点上而不会丢失。Flume 提供了三种级别的可靠性保障。

2）可扩展性：采用了三层架构，分别为 Agent、Collector 和 Storage，每一层均可以水平扩展。

3）可管理性：所有 Agent 和 Collector 由 Master 统一管理，这使得系统便于维护。用户可以在 Master 上查看各个数据源或者数据流的执行情况，且可以对各个数据源进行配置和动态加载。

4）功能可扩展性：用户可以根据需要添加自己的 Agent、Collector 或者 Storage。

3. Flume 的核心概念

1）Event：事件，是 Flume 中的基本数据单元。它由一个头部和一个主体组成，其中头部包含元数据，主体包含实际数据。事件可以从数据源中收集，经过通道传输，最终到达接收器。

2）Client：客户端，是 Flume 的一个组件，用于将数据发送到 Flume Agent。它可以是一个简单的脚本或应用程序，也可以是一个专门的数据收集器。

3）Agent：代理，是 Flume 的一个进程，用于接收来自客户端的数据，并将其传输到目标位置。它由一个或多个 Source、Channel 和 Sink 组成，可以根据需要进行配置。

因此，Flume 的工作流程是从数据源收集数据，将其封装为事件，通过客户端发送到代理，代理将事件存储在通道中，最终将其发送到接收器。这使得 Flume 成为一个非常有用的工具，可以轻松地将数据从一个地方传输到另一个地方。

7.2 Flume 构成和工作原理

7.2.1 Flume 构成

Flume 是一个分布式、可靠、可扩展的系统，用于高效地收集、聚合和传输大量日志数据。它由 Source、Channel、Sink 三部分组成，如图 7-1 所示。

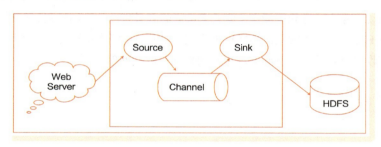

• 图 7-1 Flume 架构

Flume 的结构以及它们的功能如下。

1）Source：数据源，用于从数据源收集数据并将其传输到 Flume 中。Flume 提供了多种 Source 插件，如 ExecSource、AvroSource、NetcatSource 等，可以根据实际需求选择合适的插件。

2）Channel：通道，用于在 Flume 中存储数据。Flume 提供了多种 Channel 插件，如 MemoryChannel、JDBCChannel、KafkaChannel 等，可以根据实际需求选择合适的插件。

3）Sink：接收器，用于将数据从 Flume 发送到目标位置。Flume 提供了多种 Sink 插件，如 HDF-SEventSink、KafkaSink、LoggerSink 等，可以根据实际需求选择合适的插件。

7.2.2 Flume 工作原理

Flume 是一个分布式、可靠、高可用的海量日志采集、聚合和传输系统。它的工作原理可以简单概括为以下几个步骤。

1）采集数据：Flume 通过 Source 组件从数据源（如日志文件、网络端口等）中采集数据，并将数据存储到内存缓冲区中。

2）聚合数据：Flume 通过 Channel 组件将内存缓冲区中的数据聚合成批量数据，并将批量数据存储到磁盘缓冲区中。

3）传输数据：Flume 通过 Sink 组件将磁盘缓冲区中的数据传输到目的地（如 HDFS、Kafka 等）。

4）处理数据：Flume 可以通过 Interceptor 组件对数据进行处理（如过滤、转换等）。

7.3 Flume 实战

7.3.1 Flume 监听目录实战

在本实战中，我们将介绍如何使用 Flume 监听目录，并通过实际操作来掌握其核心概念和配置方法。下面是 Flume 监听目录的配置，在 /opt/module/flum/conf 下创建配置文件 flume-hdfs.conf。

```
#1.定义 agent 的名字 a1
a1.sources = r1
a1.sinks = k1
a1.channels = c1

#2.定义 Source
a1.sources.r1.type = exec
a1.sources.r1.command = tail -F /opt/jareny
a1.sources.r1.shell = /bin/bash -c

#3.定义 sink
a1.sinks.k1.type = hdfs
a1.sinks.k1.hdfs.path = hdfs://Linux-1:9000/flume/% H
#上传文件的前缀
a1.sinks.k1.hdfs.filePrefix = Andy-
#是否按照时间滚动文件夹
a1.sinks.k1.hdfs.round = true
#多少时间单位创建一个新的文件夹
a1.sinks.k1.hdfs.roundValue = 1
#重新定义时间单位
a1.sinks.k1.hdfs.roundUnit = hour
#是否使用本地时间戳
a1.sinks.k1.hdfs.useLocalTimeStamp = true
#积攒多少个 Event 才 flush 到 HDFS 一次
a1.sinks.k1.hdfs.batchSize = 1000
#设置文件类型,可支持压缩
a1.sinks.k1.hdfs.fileType = DataStream
#多久生成一个新的文件
a1.sinks.k1.hdfs.rollInterval = 600
#设置每个文件的滚动大小
a1.sinks.k1.hdfs.rollSize = 134217700
#文件的滚动与 Event 数量无关
a1.sinks.k1.hdfs.rollCount = 0
#最小副本数
```

```
a1.sinks.k1.hdfs.minBlockReplicas = 1

#4.定义 Channel
a1.channels.c1.type = memory
a1.channels.c1.capacity = 1000
a1.channels.c1.transactionCapacity = 100

#5.连接
a1.sources.r1.channels = c1
a1.sinks.k1.channel = c1
```

7.3.2 Flume 一对多实战

在本实战中，我们将介绍如何使用 Flume 监听目录一对多的配置。

1. 创建配置文件

Flume 监听的配置，在/opt/module/flume/conf 创建配置文件 flume1.conf，用于监控某文件的变动，同时产生两个 Channel 和两个 Sink 分别输送给 flume2 和 flume3，具体配置如下。

```
# Name the components on this agent
a1.sources = r1
a1.sinks = k1 k2
a1.channels = c1 c2
# 将数据流复制给多个 channel
a1.sources.r1.selector.type = replicating

# Describe/configure the source
a1.sources.r1.type = exec
a1.sources.r1.command = tail -F /opt/jareny
a1.sources.r1.shell = /bin/bash -c

# Describe the sink
a1.sinks.k1.type = avro
a1.sinks.k1.hostname = Linux-1
a1.sinks.k1.port = 4141

a1.sinks.k2.type = avro
a1.sinks.k2.hostname =  Linux-1
a1.sinks.k2.port = 4142

# Describe the channel
a1.channels.c1.type = memory
a1.channels.c1.capacity = 1000
a1.channels.c1.transactionCapacity = 100

a1.channels.c2.type = memory
```

```
a1.channels.c2.capacity = 1000
a1.channels.c2.transactionCapacity = 100

# Bind the source and sink to the channel
a1.sources.r1.channels = c1 c2
a1.sinks.k1.channel = c1
a1.sinks.k2.channel = c2
```

2. 创建多配置

1）创建 flume2.conf，用于接收 flume1 的 Event，同时产生一个 Channel 和一个 Sink，将数据输送给 HDFS。

```
# 1 agent
a2.sources = r1
a2.sinks = k1
a2.channels = c1

# 2 source
a2.sources.r1.type = avro
a2.sources.r1.bind = Linux-2
a2.sources.r1.port = 4141

# 3 sink
a2.sinks.k1.type = hdfs
a2.sinks.k1.hdfs.path = hdfs://bigdata111:9000/flume2/%H
#上传文件的前缀
a2.sinks.k1.hdfs.filePrefix = flume2-
#是否按照时间滚动文件夹
a2.sinks.k1.hdfs.round = true
#多少时间单位创建一个新的文件夹
a2.sinks.k1.hdfs.roundValue = 1
#重新定义时间单位
a2.sinks.k1.hdfs.roundUnit = hour
#是否使用本地时间戳
a2.sinks.k1.hdfs.useLocalTimeStamp = true
#积攒多少个 Event 才 flush 到 HDFS 一次
a2.sinks.k1.hdfs.batchSize = 100
#设置文件类型,可支持压缩
a2.sinks.k1.hdfs.fileType = DataStream
#多久生成一个新的文件
a2.sinks.k1.hdfs.rollInterval = 600
#设置每个文件的滚动大小大概是 128MB
a2.sinks.k1.hdfs.rollSize = 134217700
#文件的滚动与 Event 数量无关
a2.sinks.k1.hdfs.rollCount = 0
#最小副本数
```

```
a2.sinks.k1.hdfs.minBlockReplicas = 1

# 4 channel
a2.channels.c1.type = memory
a2.channels.c1.capacity = 1000
a2.channels.c1.transactionCapacity = 100

#5 Bind
a2.sources.r1.channels = c1
a2.sinks.k1.channel = c1
```

2）创建 flume3.conf，用于接收 flume1 的 Event，同时产生一个 Channel 和一个 Sink，将数据输送给本地目录。

```
# Name the components on this agent
a3.sources = r1
a3.sinks = k1
a3.channels = c1

# Describe/configure the source
a3.sources.r1.type = avro
a3.sources.r1.bind = bigdata111
a3.sources.r1.port = 4142

# Describe the sink
a3.sinks.k1.type = file_roll
#备注:此处的文件夹需要先创建好
a3.sinks.k1.sink.directory = /opt/flume3

# Describe the channel
a3.channels.c1.type = memory
a3.channels.c1.capacity = 1000
a3.channels.c1.transactionCapacity = 100

# Bind the source and sink to the channel
a3.sources.r1.channels = c1
a3.sinks.k1.channel = c1
```

注意：输出的本地目录必须是已经存在的目录，如果该目录不存在，并不会创建新的目录。

7.3.3 Flume 拦截器实战

1. 创建配置

Flume 正则过滤拦截器的配置 filter.conf，具体配置如下。

```
a1.sources = r1
a1.sinks = k1
```

```
a1.channels = c1

a1.sources.r1.type = exec
a1.sources.r1.channels = c1
a1.sources.r1.command = tail -F /opt/jareny
a1.sources.r1.interceptors = i1
a1.sources.r1.interceptors.i1.type = regex_filter
a1.sources.r1.interceptors.i1.regex = ^A.*
#如果 excludeEvents 设为 false,表示过滤掉不是以 A 开头的 events。
#如果 excludeEvents 设为 true,则表示过滤掉以 A 开头的 events。
a1.sources.r1.interceptors.i1.excludeEvents = true

a1.sinks.k1.type = logger

a1.channels.c1.type = memory
a1.channels.c1.capacity = 1000
a1.channels.c1.transactionCapacity = 100

a1.sources.r1.channels = c1
a1.sinks.k1.channel = c1
```

2. 正则抽取拦截器

设置正则抽取拦截器的配置 **extractor.conf**，具体配置如下。

```
a1.sources = r1
a1.sinks = k1
a1.channels = c1

a1.sources.r1.type = exec
a1.sources.r1.channels = c1
a1.sources.r1.command = tail -F /opt/jareny
a1.sources.r1.interceptors = i1
a1.sources.r1.interceptors.i1.type = regex_extractor
a1.sources.r1.interceptors.i1.regex = hostname is (.* ?) ip is (.* )
a1.sources.r1.interceptors.i1.serializers = s1 s2
a1.sources.r1.interceptors.i1.serializers.s1.name = cookieid
a1.sources.r1.interceptors.i1.serializers.s2.name = ip

a1.sinks.k1.type = logger

a1.channels.c1.type = memory
a1.channels.c1.capacity = 1000
a1.channels.c1.transactionCapacity = 100

a1.sources.r1.channels = c1
a1.sinks.k1.channel = c1
```

7.3.4 Flume 采集数据到 HDFS

在以下的实战操作中,我们将深入探讨 Flume 的数据采集和传输机制,并通过实际操作来掌握如何配置 Flume 以将数据写入 HDFS。

Flume 组件的选型,Source 为 exec,Sink 为 hdfs,Channel 为 memory。Flume 采集数据配置,在 /opt/module/flume/conf/ 目录下创建 Flueme 采集数据到 HDFS 的配置文件 exec-hdfs.conf,具体配置如下。

```
#定义三大组件的名称,agent 的自定义为 a1
a1.sources = r1
a1.sinks = k1
a1.channels = c1

# 配置 source 组件
a1.sources.r1.type = exec
a1.sources.r1.command = tail -F /opt/jareny/bigdata/logs/nginx-hdfs.log
a1.sources.r1.shell = /bin/bash -c

# 配置 sink 组件
# sink 的类型
a1.sinks.k1.type = hdfs
# 输出的 hdfs 路径
a1.sinks.k1.hdfs.path = hdfs://Linux-1:9000/jareny/bigdata/flume/exec/log/%y-%m-%d/%H-%M
# 生成文件的前缀(收集到的文件会被重命名成随机字符串)
a1.sinks.k1.hdfs.filePrefix = jareny-bigdata-flum-

## 控制生成目录的规则
# 是否自动生成目录
a1.sinks.k1.hdfs.round = true
# 自动生成目录的时间值
a1.sinks.k1.hdfs.roundValue = 1
# 自动生成目录的时间单位
a1.sinks.k1.hdfs.roundUnit = hour
#控制文件的滚动频率
#时间维度
a1.sinks.k1.hdfs.rollInterval = 30
#文件大小维度
a1.sinks.k1.hdfs.rollSize = 13417700
#event 数量维度
a1.sinks.k1.hdfs.rollCount = 0
# 当文件达到多大时,刷新到 hdfs,单位 MB
a1.sinks.k1.hdfs.batchSize = 1000
# 是否使用本地的时间戳取时间
```

```
a1.sinks.k1.hdfs.useLocalTimeStamp = true
#生成的文件类型，默认是Sequencefile，可用DataStream，则为普通文本
a1.sinks.k1.hdfs.fileType = DataStream

# channel 组件配置
# channel 将数据存储的位置一般为memory/File
a1.channels.c1.type = memory
a1.channels.c1.capacity = 1000
a1.channels.c1.transactionCapacity = 100

# 绑定 source、channel 和 sink 之间的连接
a1.sources.r1.channels = c1
a1.sinks.k1.channel = c1
```

7.3.5 Kafka 对接 Flume 实战

在以下的实战操作中，我们将深入探讨 Kafka 与 Flume 之间的集成原理，并通过实际操作来掌握如何配置 Kafka 与 Flume 进行数据传输。

Flume 组件的选型，Source 为 exec，Sink 为 kafka，Channel 为 memory。Flume 配置，在 /opt/module/flume/conf/ 目录下创建 Flume 采集数据的配置文件 exec-kafka.conf，具体配置如下。

```
##定义三大组件的名称，agent 的自定义为 a1
agent.sources = execSource
agent.channels = memoryChannel
agent.sinks = kafkasink

# 配置 source 组件
agent.sources.execSource.type = exec
# 下面这个路径是需要收集日志的绝对路径，改为自己的日志目录
agent.sources.execSource.command = tail -F /opt/jareny/bigdata/logs/nginx-kafka.log

# 配置 sink 组件
# sink 的类型为 kafka 接收器
agent.sinks.kafkasink.type = org.apache.flume.sink.kafka.KafkaSink
# 定义 kafka 的 topic
agent.sinks.kafkasink.topic = flume-kafka-producer
# 设置 kafka 的 broker 集群地址和端口号
agent.sinks.kafkasink.brokerList = Linux-1:9092,Linux-2:9092,Linux-3:9092
# 定义 kafka 的消息发送(ACK)机制
agent.sinks.kafkasink.requiredAcks = 1
# 定义 kafka 的单位 MB
agent.sinks.kafkasink.batchSize = 20

# channel 组件配置
```

```
# channel 将数据存储的位置一般为 memory/File
agent.channels.memoryChannel.type = memory
agent.channels.memoryChannel.capacity = 10000
agent.channels.memoryChannel.transactionCapacity = 100

# 绑定 source、channel 和 sink 之间的连接
agent.sources.execSource.channels = memoryChannel
agent.sinks.kafkasink.channel = memoryChannel
```

第 8 章 发布订阅消息系统 Kafka

本章内容介绍 Kafka 的基本概念，如主题（topic）、分区（partition）、副本（replica）等；Kafka 的工作原理，如消息发送确认机制、消息同步机制、消费机制等；Kafka 的生产者和消费者的使用方法，如何使用 Kafka 发送和接收消息；Kafka 的配置和使用，如何配置 Kafka 集群，如何启动和停止 Kafka 集群。Kafka 的应用案例，包括日志收集、数据流处理等场景。

8.1 Kafka 简介

1. Kafka 概述

Kafka 是一种分布式流处理平台，它是一种高吞吐量、低延迟的消息传递系统，可以处理大量的实时数据流。

Kafka 的核心概念是消息（message），它是一种键值对（key-value），包含了消息的内容和元数据。Kafka 将消息存储在一个或多个分区（Partition）中，每个分区都是一个有序的、不可变的消息序列。

Kafka 支持多个生产者（Producer）和多个消费者（Consumer），可以实现高效的消息传递和处理。Kafka 还支持消息的持久化存储和数据的复制（Replication），可以保证数据的可靠性和高可用性。Kafka 可以与 Hadoop、Storm、Spark 等大数据处理框架集成，可以方便地进行实时数据处理和分析。总之，Kafka 是一种高性能、可靠、可扩展的分布式流处理平台，可以满足各种实时数据处理和分析需求。

2. Kafka 的组件

Kafka 由多个组件组成，每个组件都有不同的作用。Kafka 存储的消息来自 Producer 生产者的进程，数据从而可以被发布到不同的 Topic 主题中的不同 Partition 分区。在一个分区内，这些消息被索引并连同时间戳存储在一起，它可以被 Consumer 消费者的进程从分区订阅消息。各个组件的关系和作用，如图 8-1 所示。

3. Kafka 的组件以及作用

1）Broker（节点）：Kafka 集群中的每个节点都是一个 Broker，它负责存储和处理消息。每个

Broker 可以处理多个分区，每个分区可以有多个副本（Replica）。Broker 之间可以进行数据的复制（Replication），以保证数据的可靠性和高可用性。

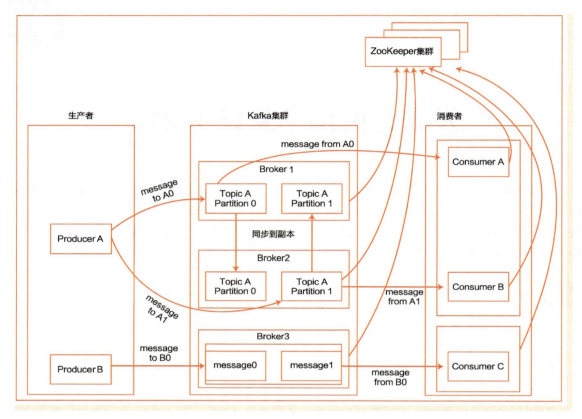

● 图 8-1　Kafka 架构图

2）Topic（主题）：Topic 是消息的逻辑分类，每个 Topic 包含一个或多个分区。生产者可以将消息发送到指定的 Topic，消费者可以从指定的 Topic 中读取消息。Topic 可以动态地创建和删除，也可以进行分区和副本的调整。

3）Partition（分片）：Partition 是 Topic 的物理分片，每个 Partition 包含一个有序的、不可变的消息序列。每个 Partition 可以有多个副本（Replica），副本之间可以进行数据的复制（Replication）。Partition 的数量和大小可以根据实际需求进行调整。

4）Producer（生产者）：Producer 是消息的生产者，它将消息发送到指定的 Topic 中。Producer 可以指定消息的键（key）和值（value），也可以指定消息的分区（Partition）。Producer 可以进行消息的批量发送和异步发送，以提高吞吐量和性能。

5）Consumer（消费者）：Consumer 是消息的消费者，它从指定的 Topic 中读取消息。Consumer 可以指定读取的位置（Offset），也可以指定读取的分区（Partition）和消费者组（Group）。Consumer 可以进行消息的批量读取和异步读取，以提高吞吐量和性能。

6）Consumer Group（消费者组）：Consumer Group 是一组消费者的集合，它们共同消费一个或多个 Topic 中的消息。每个消费者只能消费一个分区，但是一个分区可以被多个消费者共同消费。Consumer Group 可以进行动态扩展和缩减，以适应实际的负载和需求。

7）ZooKeeper（协调者）：ZooKeeper 是 Kafka 集群的协调服务，它负责管理 Broker、Topic、Partition、Producer、Consumer 等信息。ZooKeeper 可以进行数据的存储和同步，以保证 Kafka 集群的一致性和可靠性。

4. Kafka 的应用场景

Kafka 是一种分布式消息队列系统，被广泛应用于实时数据流处理、日志收集和事件驱动型微服务等领域。在本文中，我们将介绍 Kafka 的一些典型应用场景。

1）实时数据处理：Kafka 可以处理实时数据流，例如从传感器、日志、社交媒体等收集的数据。它可以对数据进行过滤、转换、聚合等操作，并将结果输出到数据库、文件系统或其他数据存储系统中。

2）日志收集：Kafka 可以用于日志收集任务，例如从多个服务器上收集日志数据，并将其存储到中心化的日志存储系统中。它可以处理大量的日志数据，并支持多个消费者同时消费数据。

3）流式处理：Kafka 可以用于流式处理任务，例如对实时数据流进行处理和分析。它可以将数据流分成多个分区，并将分区分配给不同的消费者进行处理。

4）消息队列：Kafka 可以用作消息队列，例如将消息从一个应用程序传递到另一个应用程序。它可以保证消息的顺序性和可靠性，并支持多个消费者同时消费数据。

5）数据缓存：Kafka 可以用作数据缓存，例如将数据缓存在内存中，以提高数据访问速度。它可以支持高并发的读写操作，并支持数据的持久化存储。

总之，Kafka 是一个非常强大的分布式流处理平台，可以应用于各种实时数据处理和消息传递任务。

8.2 Kafka 的消息生产者

8.2.1 Kafka 生产者的运行流程

本文我们探讨 Kafka 生产者的运行流程，包括消息的发送、序列化和存储等环节。你将了解如何使用 Kafka 生产者来发送数据到 Kafka，并了解其内部的消息处理机制。Kafka 生产者的运行流程，如图 8-2 所示。

1. Kafka 生产者的运行流程

1）一条消息过来首先会被封装成为一个 ProducerRecord 对象。

2）接下来要对这个对象进行序列化，因为 Kafka 的消息需要从客户端传到服务端，涉及网络传输，所以需要实现序列。Kafka 提供了默认的序列化机制，也支持自定义序列化。

3）消息序列化完成后，对消息要进行分区，分区的时候需要获取集群的元数据。分区的这个过程很关键，因为这个时候就决定了我们的这条消息会被发送到 Kafka 服务端的哪个主题的哪个分区。

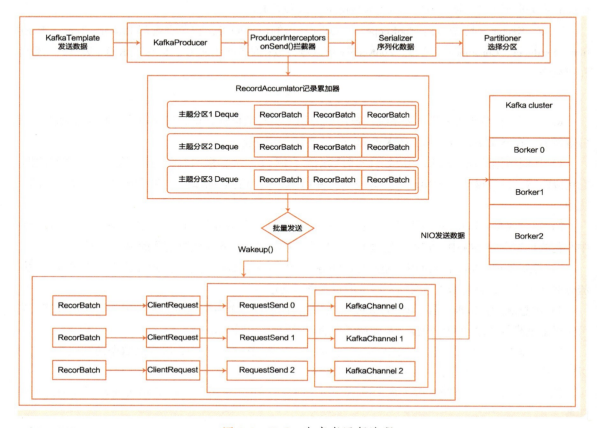

• 图 8-2　Kafka 生产者运行流程

4）分好区的消息不是直接被发送到服务端，而是放入了生产者的一个缓存里面。在这个缓存里面，多条消息会被封装成为一个批次（batch），默认一个批次的大小是 16KB。

5）Sender 线程启动以后会从缓存里面获取可以发送的批次。

6）Sender 线程把一个个批次发送到服务端。

2. Kafka 生产者消息写入方式

Producer 采用推（Push）模式将消息发布到 Broker，每条消息都被追加（Append）到分区（Partition）中，属于顺序写磁盘（顺序写磁盘效率比随机写内存要高，保障 Kafka 吞吐率）。

▶▶ 8.2.2　Kafka 生产者分区

Kafka 消息被发送到一个 Topic，其本质就是一个目录，而 Topic 由一些 Partition Logs（分区日志）组成。

1. 分区原因

1）方便在集群中扩展，每个 Partition 可以通过调整以适应它所在的机器，而一个 Topic 又可以由多个 Partition 组成，因此整个集群就可以适应任意大小的数据了。

2）可以提高并发，因为可以以 Partition 为单位读写了。

2. 分区原则

我们需要将 Producer 发送的数据封装成一个 ProducerRecord 对象。

1）指明 Partition 的情况下，直接将指明的值作为 Partition 值。

2）没有指明 Partition 值但有 key 的情况下，将 key 的 hash 值与 Topic 的 Partition 数进行取余 Partition 值。

3）既没有 Partition 值又没有 key 值的情况下，第一次调用时随机生成一个整数（后面每次调用在这个整数上自增），将这个值 Topic 可用的 Partition 总数取余得到 Partition 值，也就是常说的 round-robin 算法。

8.2.3 副本的同步复制和异步复制

Kafka 动态维护了一个同步状态的副本的集合，简称 ISR，这个集合中的节点都是和 Leader 保持高度一致的，任何一条消息只有被这个集合中的每个节点读取并追加到日志中，才会向外部通知这个消息已经被提交。

1. 同步机制

1）通过配置 producer.type 的值来确定是异步还是同步，默认为同步。

异步提供了批量发送的功能。当满足以下条件之一时就触发发送。

- batch.num.messages 异步发送是每次批量发送的条目。
- queue.buffering.max.ms 异步发送时发送时间间隔，单位是毫秒。

2）Producer 的这种在内存缓存消息，当累计达到阈值时批量发送请求，小数据 IO 太多，会拖慢整体的网络延迟，批量延迟发送事实上提升了网络效率。但是如果是在达到阈值前，Producer 不可用了，缓存的数据将会丢失。

3）异步发送消息的实现很简单，客户端消息发送过来以后，先放入一个队列中，然后就返回了。Producer 再开启一个线程（ProducerSendThread）不断从队列中取出消息，然后调用同步发送消息的接口将消息发送给 Broker。

2. 同步复制流程

Kafka 的同步复制流程如图 8-3 所示。

1）Producer 连接 ZooKeeper 识别 Leader。

2）Producer 向 Leader 发送消息。

3）Leader 收到消息写入到本地 log。

4）Follower 从 Leader 拉取消息。

5）Follower 向本地写入 log。

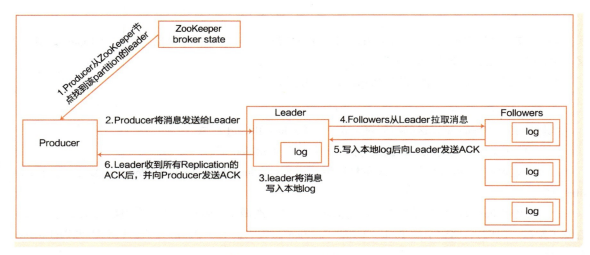

● 图 8-3 Kafka 同步复制流程

6）Follower 向 Leader 发送 ACK 消息。
7）Leader 收到所有 Follower 的 ACK 消息。
8）Leader 向 Producer 回传 ACK。

3. 异步复制流程

Kafka 中 Producer 异步发送消息是基于同步发送消息的接口来实现的，异步发送消息的实现很简单，客户端消息发送过来以后，先放入一个 BlockingQueue 队列中，然后就返回了。Producer 再开启一个线程（ProducerSendThread）不断从队列中取出消息，然后调用同步发送消息的接口将消息发送给 Broker。

8.2.4 Kafka 消息发送确认机制

为保证 Producer 发送的数据，能可靠地发送到指定的 Topic，Topic 的每个 Partition 收到 Producer 数据后，都需要向 Producer 发送 ACK（acknowledgement 确认收到），如果 Producer 收到 ACK，就会进行下一轮的发送，否则重新发送数据。

可以通过 request.required.acks 参数来设置数据可靠性的级别。

- acks=0，Producer 不等待来自 Broker 同步完成的确认就继续发送下一条（批）消息。提供最低的延迟但最弱的耐久性保证，因为其没有任何确认机制。acks 值为 0 会得到最大的系统吞吐量。
- acks=1，Producer 在 Leader 已成功收到数据并得到确认后发送下一条消息。等待 Leader 的确认后就返回，而不管 Partion 的 Follower 是否已经完成。
- acks=-1，Producer 在所有 Follower 副本确认接收到数据后才算一次发送完成。此选项提供最好的数据可靠性，只要有一个同步副本保持存活，Kafka 保证信息将不会丢失。

8.3 Kafka 的 Broker 保存消息

8.3.1 存储方式与策略

物理上把 Topic 分成一个或多个 Partition，可以通过配置设定 Partition，每个 Partition 物理上对应一个文件夹。

根据 Kafka 的设计，消息存储在 Broker 中，每个 Broker 都是一个独立的节点，可以存储多个分区。每个分区都是一个有序的、不可变的消息序列，可以持久化到磁盘上。Kafka 使用了一种基于日志的存储方式，即将消息追加到日志文件的末尾。这种存储方式具有高效、可靠、可扩展等优点。

Kafka Broker 的消息存储策略包括两种：消息保留策略和消息清理策略。消息保留策略指定了 Kafka Broker 中消息的保留时间，可以根据时间、大小或数量等条件来保留消息。消息清理策略指定了 Kafka Broker 中消息的清理方式，可以根据时间、大小或数量等条件来清理消息。Kafka 提供了多种消息保留策略和消息清理策略，可以根据实际需求进行配置。

8.3.2 Topic 创建与删除

在 Kafka 中，每条发送到 Kafka 集群的消息都有一个类别，这个类别叫作 Topic。Topic 是一个存储消息的逻辑概念，可认为是一个消息集合。每个 Topic 可以有多个生产者向它发送消息，也可以有多个消费者去消费其中的消息。

1. Kafka 创建 Topic 流程

1）确定 Topic 名称、分区数和副本数。
2）使用命令行工具创建 Topic，指定 ZooKeeper 地址、分区数、副本数和 Topic 名称。
3）Kafka Broker 接收到创建 Topic 的请求后，会在 ZooKeeper 中创建一个对应的节点，表示该 Topic 已经存在。
4）Kafka Broker 会为每个分区创建一个对应的日志文件，用于存储该分区的消息。
5）Kafka Broker 会将每个分区的副本分配到不同的节点上，以保证数据的可靠性和高可用性。

2. Kafka 删除 Topic 流程

1）使用命令行工具删除 Topic，指定 ZooKeeper 地址和 Topic 名称。
2）Kafka Broker 接收到删除 Topic 的请求后，会在 ZooKeeper 中删除对应的节点，表示该 Topic 已经不存在。
3）Kafka Broker 会删除该 Topic 对应的所有分区的日志文件，以及所有副本的日志文件。
4）Kafka Broker 会将该 Topic 对应的所有分区的副本从所有节点上删除。

8.4 Kafka 的消息消费者

Kafka 的消费端以消费者组的方式进行消费，一个消费者组中存在多个消费者，但是它们的

groupId 是一致的,当一个消费者组订阅了一个 Topic 主题时,组内的消费者协调一起消费 Topic 中的所有分区,且一个分区只能由一个消费者消费,消费端可以通过设置 TopicPartition 来指定消费的分区。消费模型如图 8-4 所示。

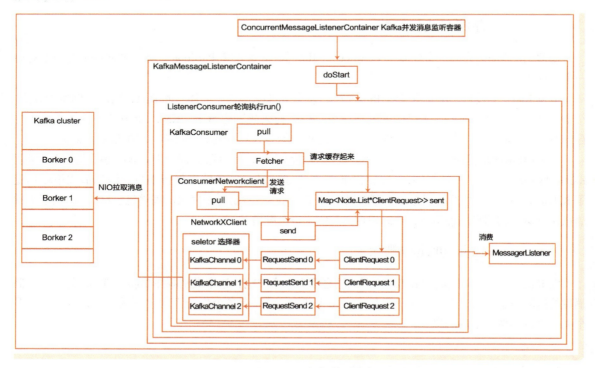

● 图 8-4 消费模型

8.4.1 消费机制

Kafka 消费机制是 Kafka 分布式流处理系统的重要组成部分,它提供了高效、可靠和可扩展的数据消费功能。Kafka 消费机制的基本原理如下。

1) 一个消息只能被同一个消费组的某个消费者消费,消费后不删除消息只是自己消费消息的 offset+1,并不会删除消息,可能消息还要被其他消费组消费,可配置清除默认 7 天或基于大小来清除旧消息。

2) 可以手动指定那些 Partition 由组内哪个消费者消费,不指定则首次某个消费者消费某个 Partition 后不再改变,如果消费者 offline 或 online,会通过 Rebalance 力求负载均衡。

3) 每个 Partition 都维护每个消费组的 offset,故只能保证同一个 Partition 的数据是顺序消费的,不能保证整个 Topic 的消息顺序消费,这个顺序指的是 Partititon 分区内顺序,不是消息发送的顺序,即使同一个 Partition 已不能绝对保证发送顺序与消费顺序一致。

4) 如果 Consumer 个数大于 Partition 数,则部分 Consumer 一直消费不到数据,处于空闲状态浪费资源,故分区数应大于等于消费者数。

8.4.2 消费者组

1. 消费者组概述

消费者是以 Consumer Group 消费者组的方式工作，由一个或者多个消费者组成一个组，共同消费一个 Topic。每个分区在同一时间只能由 Group 中的一个消费者读取，但是多个 Group 可以同时消费这个 Partition。有一个由三个消费者组成的 Group，有一个消费者读取主题中的两个分区，另外两个分别读取一个分区。某个消费者读取某个分区，也可以叫作某个消费者是某个分区的拥有者。

在这种情况下，消费者可以通过水平扩展的方式同时读取大量的消息。另外，如果一个消费者失败了，那么其他的 Group 成员会自动负载均衡读取之前失败的消费者读取的分区。

2. 消费方式

在消费者组消费中，消费者采用 pull 方式消费。具体来说，消费者会定期向 Kafka broker 发送拉取请求，获取分配给自己的分区中的消息。消费者可以控制拉取请求的大小和频率，以适应不同的业务场景和网络环境。消费者还可以控制拉取请求的超时时间，以避免长时间等待而导致的性能问题。

3. 消费分配策略

Kafka 的消费者组分区分配策略是基于消费者组中的消费者数量和 Topic 的分区数量来确定的。具体来说，Kafka 会将每个 Topic 分为多个分区，每个分区会分配给消费者组中的一个消费者进行消费。如果消费者组中的消费者数量小于 Topic 的分区数量，那么有些消费者将会被分配到多个分区上进行消费，以实现负载均衡。如果消费者组中的消费者数量大于 Topic 的分区数量，那么有些消费者将会处于空闲状态，等待新的分区加入。Kafka 的消费者组分区分配策略是动态的，可以根据消费者组中的消费者数量和 Topic 的分区数量进行自适应调整，以实现高吞吐量和可靠性。

Kafka 有两种分配策略，RoundRobin：即按轮询划分；Range：按范围划分。Kafka 默认采用按范围分配策略。

4. offset 的维护

由于 Consumer 在消费过程中可能会出现断电宕机等故障，Consumer 恢复后，需要从故障前的位置继续消费，所以 Consumer 需要实时记录自己消费到了哪个 offset，以便故障恢复后继续消费。

8.5 Kafka 的存储机制

Kafka 是为了解决大数据的实时日志流而生的，日志流的特点主要包括：数据实时产生和海量数据存储与处理。

对于 Kafka 来说，它主要用来处理海量数据流，直接采用顺序追加写日志的方式就可以满足 Kafka 对于百万 TPS 写入效率的要求；对于高效查询这些日志，Kafka 采用了稀疏哈希索引的方式，如图 8-5 所示。

Topic 是一个消息队列，每条发送到 Kafka 集群的消息都有一个 Topic。物理上来说，不同的 Topic

的消息是分开存储的，每个 Topic 可以有多个生产者向它发送消息，也可以有多个消费者去消费其中的消息，每个 Topic 可以拆分为多个 Partition 分区，分区内部会给消息分配一个 offset 来表示消息的顺序所在，这就是 Kafka 的存储机制。

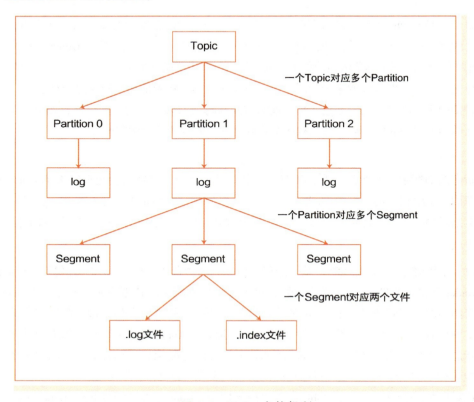

● 图 8-5 Kafka 存储机制

8.5.1 Kafka 主题 Topic

在 Kafka 中的 Topic 是一个存储消息的逻辑概念，可以理解为是一个消息的集合。每条发送到 Kafka 集群的消息都会自带一个类别，表明要将消息发送到哪个 Topic 上。同类消息发送到同一个 Topic 下面，如图 8-6 所示。

在存储方面，不同的 Topic 的消息是分开存储的，每个 Topic 可以有多个生产者向它发送消息，也可以有多个消费者去消费同一个 Topic 中的消息。Topic 是 Kafka 数据写入操作的基本单元，可以指定副本。Kafka Topic 的基本描述如下。

1) 一个 Topic 包含一个或多个 Partition，建 Topic 的时候可以手动指定 Partition 个数，个数与服务器个数相当。

2) 每条消息属于且仅属于一个 Topic。

3) Producer 发布数据时，必须指定将该消息发布到哪个 Topic。

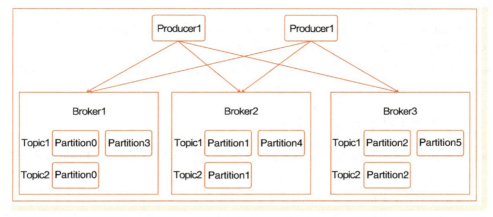

- 图 8-6 Kafka 的 Topic 示意图

4）Consumer 订阅消息时，也必须指定订阅哪个 Topic 的信息。

通过将消息按照 Topic 分类，Kafka 提供了一种高效、可靠和可扩展的消息传递机制。应用程序可以根据需要订阅不同的 Topic，实现灵活的消息处理和分发。同时，Kafka 的高可用性和分布式存储特性也确保了消息的安全性和可靠性。

8.5.2 Kafka 分片 Partition

Kafka 可以将主题 Topic 划分为多个分区 Partition 中。在 Kafka 存在多副本的情况下，会尽量把多个副本分配到不同的 broker 上。Kafka 会为 Partition 选出一个 Leader，之后所有该 Partition 的请求，实际操作的都是 Leader，然后同步到其他的 Follower。当一个 broker 宕机后，所有 Leader 在该 broker 上的 Partition 都会重新选举，选出一个 Leader，如图 8-7 所示。

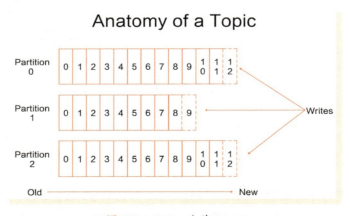

- 图 8-7 Kafka 分片 Partition

Kafka 的每个 Partition 只会在一个 broker 上，物理上每个 Partition 对应的是一个文件夹。

1）Kafka 默认使用的是 hash 进行分区，所以会出现不同的分区数据不一样的情况，但是

Partitioner 是可以覆盖的。

2) 一个 Partition 包含多个 Segment，每个 Segment 对应一个文件，Segment 可以手动指定大小，当 Segment 达到阈值时，将不再写数据，每个 Segment 都是大小相同的。

3) Segment 由多个不可变的记录组成，记录只会被 append 到 Segment 中，不会被单独删除或者修改，每个 Segment 中的 Message 数量不一定相等。

8.5.3 Kafka 日志 Segment File

Kafka 实际是用先写内存映射的文件、磁盘顺序读写的技术来提高性能的。Producer 生产的消息按照一定的分组策略被发送到 broker 中的 Partition 中时，这些消息如果在内存中放不下了，就会放在 Partition 目录下的文件中，Partition 目录名是 Topic 的名称加上一个序号。

在这个目录下有两类文件，一类是以.log 为后缀的文件，另一类是以.index 为后缀的文件，每一个.log 文件和一个.index 文件相对应，这一对文件就是一个 Segment File，其中的.log 文件就是数据文件，里面存放的就是 Message，而.index 文件是索引文件。Index 文件记录了元数据信息，指向对应的数据文件中 Message 的物理偏移量。存储格式如图 8-8 所示。

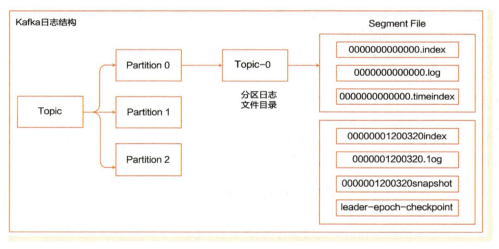

● 图 8-8 存储格式

8.6 Kafka 实战

在本实战中，我们将介绍如何使用 Kafka 发送消息，包括消息的发送、序列化和存储等环节。通过实际操作，你将了解如何使用 Kafka 生产者来发送数据到 Kafka，并了解其内部的消息处理机制。

8.6.1 Kafka 发送消息

生产者的作用是将数据发送到 Kafka 集群中的一个或多个主题。以下是一个简单的

Kafka 生产者代码示例。

```java
import org.apache.kafka.clients.producer.*;
import java.util.Properties;

public class KafkaProducerExample {
    public static void main(String[] args) {
        String topicName = "my-topic";
        String key = "my-key";
        String value = "Hello, Kafka!";

        Properties props = new Properties();
        props.put("bootstrap.servers", "localhost:9092");
        props.put("key.serializer", "org.apache.kafka.common.serialization.StringSerializer");
        props.put("value.serializer", "org.apache.kafka.common.serialization.StringSerializer");

        Producer<String, String> producer = new KafkaProducer<>(props);

        ProducerRecord<String, String> record = new ProducerRecord<>(topicName, key, value);

        producer.send(record);

        producer.close();
    }
}
```

在这个例子中，我们使用 Kafka 的 Java 客户端库来创建一个生产者。首先定义了要发送的主题名称、键和值。然后设置了一些 Kafka 生产者的属性，例如 Kafka 集群的地址和序列化器。接下来创建了一个 Kafka 生产者实例，并使用它来发送一个包含键、值和主题名称的记录。最后关闭了生产者。

接下来，让我们看一下 Kafka 消费者的代码。消费者的作用是从 Kafka 集群中的一个或多个主题中读取数据。以下是一个简单的 Kafka 消费者代码示例。

```java
import org.apache.kafka.clients.consumer.*;
import java.util.Collections;
import java.util.Properties;

public class KafkaConsumerExample {
    public static void main(String[] args) {
        String topicName = "my-topic";

        Properties props = new Properties();
        props.put("bootstrap.servers", "localhost:9092");
        props.put("key.deserializer", "org.apache.kafka.common.serialization.StringDeserializer");
```

```java
        props.put("value.deserializer", "org.apache.kafka.common.serialization.StringDe-
serializer");
        props.put("group.id", "my-group");

        KafkaConsumer<String, String> consumer = new KafkaConsumer<>(props);

        consumer.subscribe(Collections.singletonList(topicName));

        while (true) {
            ConsumerRecords<String, String> records = consumer.poll(100);

            for (ConsumerRecord<String, String> record : records) {
                System.out.printf("offset = %d, key = %s, value = %s%n", record.offset(),
record.key(), record.value());
            }
        }
    }
}
```

在这个例子中,我们使用 Kafka 的 Java 客户端库来创建一个消费者。首先定义了要读取的主题名称。然后设置了一些 Kafka 消费者的属性,例如 Kafka 集群的地址、反序列化器和消费者组 ID。接下来创建了一个 Kafka 消费者实例,并使用它来订阅一个主题。最后使用一个无限循环来轮询 Kafka 集群,读取新的记录并对它们进行处理。

8.6.2　Kafka 自定义分区发送消息

Kafka 自定义分区的生产者的代码。自定义分区的作用是将数据发送到 Kafka 集群中的一个或多个主题,并根据自定义的分区策略将数据分配到不同的分区中。以下是一个简单的 Kafka 自定义分区的生产者代码示例。

```java
import org.apache.kafka.clients.producer.*;
import org.apache.kafka.common.Cluster;
import org.apache.kafka.common.PartitionInfo;
import java.util.List;
import java.util.Map;
import java.util.Properties;

public class KafkaCustomPartitionerProducer {
    public static void main(String[] args) {
        String topicName = "my-topic";
        String key = "my-key";
        String value = "Hello, Kafka!";

        Properties props = new Properties();
        props.put("bootstrap.servers", "localhost:9092");
```

```java
        props.put("key.serializer", "org.apache.kafka.common.serialization.StringSerializer");
        props.put("value.serializer", "org.apache.kafka.common.serialization.StringSerializer");
        props.put("partitioner.class", "com.example.CustomPartitioner");

        Producer<String, String> producer = new KafkaProducer<>(props);

        ProducerRecord<String, String> record = new ProducerRecord<>(topicName, key, value);

        producer.send(record, new Callback() {
            @Override
            public void onCompletion(RecordMetadata metadata, Exception exception) {
                if (exception != null) {
                    exception.printStackTrace();
                } else {
                    System.out.printf("sent record to topic=%s partition=%d%n", metadata.topic(), metadata.partition());
                }
            }
        });

        producer.close();
    }
}
```

在这个例子中,我们使用 Kafka 的 Java 客户端库来创建一个自定义分区的生产者。首先定义了要发送的主题名称、键和值。然后设置了一些 Kafka 生产者的属性,例如 Kafka 集群的地址和序列化器。接下来指定了自定义分区策略的类名。最后创建了一个 Kafka 生产者实例,并使用它来发送一个包含键、值和主题名称的记录。在发送记录时,还提供了一个回调函数,以便在记录发送完成后进行处理。最后关闭了生产者。

接下来,让我们看一下 Kafka 自定义分区的消费者的代码。消费者的作用是从 Kafka 集群中的一个或多个主题中读取数据。以下是一个简单的 Kafka 自定义分区的消费者代码示例。

```java
import org.apache.kafka.clients.consumer.*;
import org.apache.kafka.common.TopicPartition;
import java.util.*;

public class KafkaCustomPartitionerConsumer {
    public static void main(String[] args) {
        String topicName = "my-topic";

        Properties props = new Properties();
        props.put("bootstrap.servers", "localhost:9092");
        props.put("key.deserializer", "org.apache.kafka.common.serialization.StringDeserializer");
```

```
        props.put("value.deserializer", "org.apache.kafka.common.serialization.StringDe-
serializer");
        props.put("group.id", "my-group");

        KafkaConsumer<String, String> consumer = new KafkaConsumer<>(props);

        List<PartitionInfo> partitionInfos = consumer.partitionsFor(topicName);
        List<TopicPartition> partitions = new ArrayList<>();
        if (partitionInfos != null) {
            for (PartitionInfo partitionInfo : partitionInfos) {
                partitions.add(new TopicPartition(partitionInfo.topic(), partitionInfo.
partition()));
            }
        }

        consumer.assign(partitions);

        while (true) {
            ConsumerRecords<String, String> records = consumer.poll(100);

            for (ConsumerRecord<String, String> record : records) {
                System.out.printf("offset = %d, key = %s, value = %s, partition = %d%n",
record.offset(), record.key(), record.value(), record.partition());
            }
        }
    }
}
```

在这个例子中，我们使用 Kafka 的 Java 客户端库来创建一个自定义分区的消费者。首先定义了要读取的主题名称。然后设置了一些 Kafka 消费者的属性，例如 Kafka 集群的地址、反序列化器和消费者组 ID。接下来创建了一个 Kafka 消费者实例，并使用它来获取主题的所有分区。然后将消费者分配到所有分区上。最后使用一个无限循环来轮询 Kafka 集群，读取新的记录并对它们进行处理。

8.6.3　Spring Boot 整合 Kafka 发送消息

Spring Boot 是一个快速构建应用程序的框架，而 Kafka 是一个分布式消息队列系统，可以用于大规模的数据传输和实时数据处理。在本实战中，我们将介绍如何使用 Spring Boot 整合 Kafka，并通过代码实现发送消息到 Kafka 集群。通过实际操作，你将了解如何使用 Spring Boot 和 Kafka 进行集成，并实现高效的数据传输。

1）创建一个新的项目，命名为 jareny-bigdata-kafka，并且在项目的 pom.xml 文件中添加与 Kakfa 相关的依赖，代码示例如下。

```
<?xml version="1.0" encoding="UTF-8"?>
<project xmlns="http://maven.apache.org/POM/4.0.0"
```

```xml
        xmlns:xsi="http://www.w3.org/2001/XMLSchema-instance"
        xsi:schemaLocation="http://maven.apache.org/POM/4.0.0 http://maven.apache.org/xsd/maven-4.0.0.xsd">
    <modelVersion>4.0.0</modelVersion>

    <groupId>com.it.jareny.bigdata</groupId>
    <artifactId>jareny-bigdata-kafka</artifactId>
    <version>1.0-SNAPSHOT</version>

    <parent>
        <groupId>org.springframework.boot</groupId>
        <artifactId>spring-boot-starter-parent</artifactId>
        <version>2.1.1.RELEASE</version>
        <relativePath/> <!-- lookup parent from repository -->
    </parent>

    <dependencies>
        <dependency>
            <groupId>org.springframework.boot</groupId>
            <artifactId>spring-boot-starter-web</artifactId>
        </dependency>
        <dependency>
            <groupId>org.springframework.kafka</groupId>
            <artifactId>spring-kafka</artifactId>
        </dependency>
    </dependencies>
</project>
```

2）创建项目启动类，命名为 KafkaApplication，代码示例如下。

```java
package com.it.jareny.bigdata.kafka;

import org.springframework.boot.SpringApplication;
import org.springframework.boot.autoconfigure.SpringBootApplication;

@SpringBootApplication
public class KafkaApplication {
    public static void main(String[] args) {
        SpringApplication.run(KafkaApplication.class,args);
    }
}
```

3）在项目的/src/main/resources/创建 application.properties 文件，并且在配置文件中配置该 Kafka 的相关参数，代码示例如下。

```properties
#端口
server.port=9000
```

```
###########【Kafka 集群】###########
spring.kafka.bootstrap-servers=192.168.81.111:9092,192.168.81.112:9092,192.168.81.113:9092
###########【初始化生产者配置】###########
# 重试次数,设置大于 0 的值 则客户端会将发送失败的记录重新发送
spring.kafka.producer.retries=0
# 应答级别:多少个分区副本备份完成时向生产者发送 ack 确认(可选 0、1、all/-1)
spring.kafka.producer.acks=1
# 批量大小 16KB
spring.kafka.producer.batch-size=16384
# 提交延时
spring.kafka.producer.properties.linger.ms=0
# 当生产端积累的消息达到 batch-size 或接收到消息 linger.ms 后,生产者就会将消息提交给 kafka
# linger.ms 为 0 表示每接收到一条消息就提交给 kafka,这时候 batch-size 其实就没用了

# 生产端缓冲区大小 32MB
spring.kafka.producer.buffer-memory = 33554432
# Kafka 提供的序列化和反序列化类
spring.kafka.producer.key-serializer=org.apache.kafka.common.serialization.StringSerializer
spring.kafka.producer.value-serializer=org.apache.kafka.common.serialization.StringSerializer
# 自定义分区器
# spring.kafka.producer.properties.partitioner.class=com.felix.kafka.producer.CustomizePartitioner

###########【初始化消费者配置】###########
# 默认的消费组 ID
spring.kafka.consumer.properties.group.id=defaultConsumerGroup
# 是否自动提交 offset
spring.kafka.consumer.enable-auto-commit=true
# 提交 offset 延时(接收到消息后多久提交 offset)
spring.kafka.consumer.auto.commit.interval.ms=1000
# 当 kafka 中没有初始 offset 或 offset 超出范围时将自动重置 offset
# earliest:重置为分区中最小的 offset;
# latest:重置为分区中最新的 offset(消费分区中新产生的数据);
# none:只要有一个分区不存在已提交的 offset,就抛出异常;
spring.kafka.consumer.auto-offset-reset=latest
# spring.kafka.consumer.auto-offset-reset=earliest
# 消费会话超时时间(超过这个时间 consumer 没有发送心跳,就会触发 rebalance 操作)
spring.kafka.consumer.properties.session.timeout.ms=120000
# 消费请求超时时间
spring.kafka.consumer.properties.request.timeout.ms=180000
# Kafka 提供的序列化和反序列化类
spring.kafka.consumer.key-deserializer=org.apache.kafka.common.serialization.StringDeserializer
spring.kafka.consumer.value-deserializer=org.apache.kafka.common.serialization.StringDeserializer
```

```
# 消费端监听的topic不存在时,项目启动会报错(关掉)
spring.kafka.listener.missing-topics-fatal=false
# 设置批量消费
# spring.kafka.listener.type=batch
# 批量消费每次最多消费多少条消息
spring.kafka.consumer.max-poll-records=50
```

4) 创建消息的生产者类,命名为 **KafkaProduce**,用于发送消息,代码示例如下。

```
package com.jareny.bigdata.kafka.controller;

import lombok.extern.slf4j.Slf4j;
import org.springframework.beans.factory.annotation.Autowired;
import org.springframework.kafka.core.KafkaTemplate;
import org.springframework.kafka.support.SendResult;
import org.springframework.util.concurrent.ListenableFutureCallback;
import org.springframework.web.bind.annotation.GetMapping;
import org.springframework.web.bind.annotation.PathVariable;
import org.springframework.web.bind.annotation.RequestMapping;
import org.springframework.web.bind.annotation.RestController;

@Slf4j
@RestController
@RequestMapping("/kafka")
public class KafkaProducer {
    @Autowired
    private KafkaTemplate<String, Object> kafkaTemplate;

    @GetMapping("/sendCallback/{message}")
    public void sendCallbackMessage(@PathVariable("message") String message) {
        kafkaTemplate.send("topic1", message).addCallback(success -> {
            // 消息发送到的topic
            String topic = success.getRecordMetadata().topic();
            // 消息发送到的分区
            int partition = success.getRecordMetadata().partition();
            // 消息在分区内的offset
            long offset = success.getRecordMetadata().offset();
            // 消息值
            Object value = success.getProducerRecord().value();
            log.info("kafka sendCallbackMessage 发送消息成功: topic={},partition={},offset={},value={}", topic, partition, offset, value);
        }, failure -> {
            log.info("kafka sendCallbackMessage 发送消息失败: {}", failure.getMessage());
        });
    }

    @GetMapping("/sendCallback/list/{message}")
```

```java
    public void sendCallbackByList(@PathVariable("message") String message) {
        kafkaTemplate.send("topic1", message).addCallback(new ListenableFutureCallback<SendResult<String, Object>>() {
            @Override
            public void onFailure(Throwable ex) {
                log.info("kafka sendCallbackByList 发送消息失败：{}", ex.getMessage());
            }

            @Override
            public void onSuccess(SendResult<String, Object> result) {
                log.info("kafka sendCallbackByList 发送消息成功: topic={},partition={},offset={}",
                        result.getRecordMetadata().topic(), result.getRecordMetadata().partition(),
                        result.getRecordMetadata().offset());
            }
        });
    }
}
```

5）创建消息消费者类，命名为 **KafkaConsumer**，用于接受消息，代码示例如下。

```java
package com.jareny.bigdata.kafka.controller;

import lombok.extern.slf4j.Slf4j;
import org.apache.kafka.clients.consumer.ConsumerRecord;
import org.springframework.kafka.annotation.KafkaListener;
import org.springframework.kafka.annotation.PartitionOffset;
import org.springframework.kafka.annotation.TopicPartition;
import org.springframework.stereotype.Component;
import java.util.List;

@Slf4j
@Component
public class KafkaConsumer {
    /**
     * @Title 指定 topic、partition、offset 消费
     * @Description 同时监听 topic1 和 topic2
     * 监听 topic1 的 0 号分区、topic2 的 "0 号和 1 号" 分区,指向 1 号分区的 offset 初始值为 8
     * id:消费者 ID
     * groupId:消费组 ID
     * topics:监听的 topic,可监听多个
     * topicPartitions:可配置更加详细的监听信息,可指定 topic、parition、offset 监听
     **/
    @KafkaListener(id = "consumer1", groupId = "felix-group", topicPartitions = {
            @TopicPartition(topic = "topic1", partitions = {"0"}),
            @TopicPartition(topic = "topic2", partitions = "0", partitionOffsets = @PartitionOffset(partition = "1", initialOffset = "8"))
    })
```

```java
    public void onCallbackMessage(ConsumerRecord<?, ?> record) {
        log.info("Kafka onCallbackMessage 消费: topic = {},partition = {},value = {},offset = {}", record.topic(), record.partition(), record.value(), record.offset());
    }

    /**
     * 批量消费
     **/
    @KafkaListener(id = "consumer2", groupId = "felix-group", topics = "topic1")
    public void onCallbackMessageList(List<ConsumerRecord<?, ?>> records) {
        log.info("onCallbackMessageList size=", records.size());
        for (ConsumerRecord<?, ?> record : records) {
            log.info("onCallbackMessageList value={}", record.value());
        }
    }
}
```

第9章 数据处理分析引擎 Spark

本章内容介绍 Spark 的基本概念和工作原理，包括 RDD、DAG、Spark 运行架构等。学习 Spark 的安装和配置，以及集群模式的部署。学习 Spark 的高级特性和应用场景，包括 Spark SQL、Spark Streaming、Spark MLlib 等功能。使用 Spark 编写应用案例，包括数据处理、数据分析等场景。

9.1 Spark 简介

Spark 是一个快速、通用、可扩展的分布式计算系统，最初由加州大学伯克利分校的 AMPLab 开发。它提供了高级 API，如 Spark SQL、Spark Streaming、MLlib 和 GraphX，可以轻松地处理大规模数据集和复杂的数据处理任务。Spark 的核心是 RDD（弹性分布式数据集），它是一个可并行操作的分布式数据集合，可以在内存中缓存数据，从而提高处理速度。Spark 还支持在 Hadoop 集群上运行，并与 Hadoop 的 HDFS 和 Yarn 集成。Spark 的优点包括高速处理、易于使用、灵活性和可扩展性。Spark 的架构如图 9-1 所示。

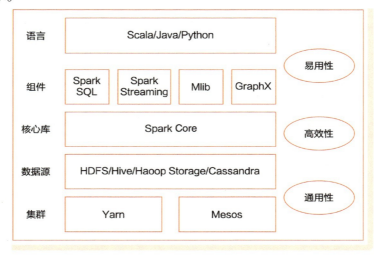

● 图 9-1 Spark 的架构体系

Spark 由 Spark Core、Spark SQL、Spark Streaming、MLlib 和 GraphX 五大部分组成。它们的功能如下。

1）Spark Core 是 Spark 的核心组件，它提供了分布式任务调度、内存管理、错误恢复、存储管理和与存储系统的交互等基本功能。Spark Core 的主要作用是提供了 RDD 的抽象，使得用户可以方便地进行分布式数据处理和分析。

2）Spark SQL 是 Spark 的一个模块，它提供了一种基于 SQL 的接口，可以让用户使用 SQL 语句来查询结构化数据。Spark SQL 支持多种数据源，包括 Hive 表、Parquet 文件、JSON 文件和 JDBC 数据源等。Spark SQL 还提供了一些高级功能，如 DataFrame 和 Dataset API，可以让用户以编程的方式操作数据。

3）Streaming 是 Spark 的一个模块，它提供了一种实时数据处理的能力，可以让用户以流式的方式处理数据。Spark Streaming 的主要作用是将实时数据流转换为一系列的小批量数据，然后使用 Spark 的批处理引擎进行处理和分析。

4）MLlib 是 Spark 的一个机器学习库，它提供了一系列的机器学习算法和工具，可以让用户以分布式的方式进行机器学习和数据挖掘。Spark MLlib 支持多种数据源，包括 Hive 表、Parquet 文件、JSON 文件和 JDBC 数据源等。Spark MLlib 还提供了一些高级功能，如 Pipeline API 和模型持久化，可以让用户以编程的方式构建和管理机器学习流程。

5）GraphX 是 Spark 的一个图处理库，它提供了一种方便、高效、灵活的方式来处理大规模图数据。Spark GraphX 支持多种图算法和操作，包括图的构建、转换、遍历、聚合和分析等。Spark GraphX 还提供了一些高级功能，如图的持久化和分布式计算，可以让用户以分布式的方式进行图处理和分析。

9.2 Spark 运行原理

9.2.1 Spark 的基本概念

本节我们将深入探讨 Spark 的基本概念，包括分布式计算、RDD 和 Spark 的架构等。你将了解如何使用 Spark 处理大规模数据集，并了解其内部的消息处理机制。学习 Spark 的运行原理，首先了解 Spark 的核心概念以及组成部分的功能，Spark 的核心概念如下。

（1）Spark 应用程序 Application

指的是用户编写的 Spark 应用程序，包含了 Driver 功能代码和分布在集群中多个节点上运行的 Executor 代码。

（2）驱动程序 Driver

Spark 中的 Driver 即运行 Application 的 Main（）函数并且创建 SparkContext，由 SparkContext 负责和 ClusterManager 通信，进行资源的申请、任务的分配和监控等。

（3）资源管理器 Cluster Manager

指的是在集群上获取资源的外部服务，常用的有：Standalone，Spark 原生的资源管理器，由

Master 负责资源的分配；Haddop Yarn，由 Yarn 中的 ResearchManager 负责资源的分配。

(4) 执行器 Executor

执行器 Executor 是 Application 运行在 Worker 节点上的一个进程，该进程负责运行 Task，并且负责将数据存在内存或者磁盘上，每个 Application 都有各自独立的一批 Executor。

(5) 计算节点

集群中任何可以运行 Application 代码的节点。在 Spark on Yarn 模式中指的就是 NodeManager 节点；在 Spark on Messos 模式中指的就是 Messos Slave 节点。

(6) 有向无环调度器 DAGScheduler

基于 DAG 划分 Stage 并以 TaskSet 的形式提交 Stage 给 TaskScheduler；负责将作业拆分成不同阶段的具有依赖关系的多批任务；计算作业和任务的依赖关系，制定调度逻辑。

(7) 任务调度器 TaskScheduler

将 Taskset 提交给 worker（集群）运行并汇报结果；负责每个具体任务的实际物理调度。

(8) 作业 Job

由一个或多个调度阶段所组成的一次计算作业；包含多个 Task 组成的并行计算，一个 Job 包含多个 RDD 及作用于相应 RDD 上的各种 Operation。

(9) 调度阶段 Stage

一个任务集对应的调度阶段；每个 Job 会被拆分成很多组 Task，每组任务被称为 Stage，也可称为 TaskSet，一个作业分为多个阶段；Stage 分成两种类型——ShuffleMapStage 和 ResultStage。

(10) 任务集 TaskSet

由一组关联的，但相互之间没有 Shuffle 依赖关系的任务所组成的任务集。一个 Stage 创建一个 TaskSet，为 Stage 的每个 RDD 分区创建一个 Task，多个 Task 封装成 TaskSet。

(11) 任务 Task

被送到某个 Executor 上的工作任务，单个分区数据集上的最小处理流程单元（单个 Stage 内部根据操作数据的分区数划分成多个 Task）。

9.2.2 Spark 运行的原理

Spark 运行架构主要由 SparkContext、Cluster Manager 和 Worker 组成，其中 Cluster Manager 负责整个集群的统一资源管理，Worker 节点中的 Executor 是应用执行的主要进程，内部含有多个 Task 线程以及内存空间。Spark 运行基本流程如图 9-2 所示。

Spark 的运行步骤如下。

1）当一个 Spark 应用被提交时，根据提交参数在相应位置创建 Driver 进程，Driver 进程根据配置参数信息初始化 SparkContext 对象，即 Spark 运行环境，由 SparkContext 负责和 Cluster Manager 的通信以及资源的申请、任务的分配和监控等。SparkContext 启动后，创建 DAG Scheduler（将 DAG 图分解成 Stage）和 Task Scheduler（提交和监控 Task）两个调度模块。

2）Driver 进程根据配置参数向 Cluster Manager 申请资源（主要是用来执行的 Executor），Cluster Manager 接收到应用（Application）的注册请求后，会使用自己的资源调度算法，在 Spark 集群的

Worker 节点上，通知 Worker 为应用启动多个 Executor。

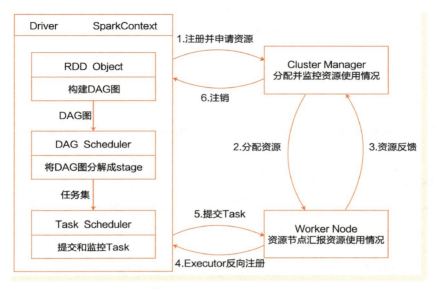

● 图 9-2 Spark 运行基本流程

3）Executor 创建后，会向 Cluster Manager 进行资源及状态的反馈，便于 Cluster Manager 对 Executor 进行状态监控，如果监控到 Executor 失败，则会立刻重新创建。

4）Executor 会向 SparkContext 反向注册申请 Task。

5）Task Scheduler 将 Task 发送给 Worker 进程中的 Executor 运行并提供应用程序代码。

6）当程序执行完毕后写入数据，Driver 向 Cluster Manager 注销申请的资源。

9.2.3 Driver 运行在 Client

Spark 的运行模式中的 Driver 运行在 Client 模式下，这意味着 Driver 程序和执行计算任务的 Executor 运行在不同的节点上。在 Client 模式下，Driver 程序运行在客户端机器上，而 Executor 则部署在集群的各个节点上。

在这种模式下，用户可以直接在客户端机器上启动 Spark 应用程序，并监视其执行情况。Driver 程序负责将任务分发到各个 Executor 上，并收集执行结果。它通过 Spark 的 API 与 Spark 集群交互，将作业提交给集群管理器并监视其执行情况。

当 Spark 应用程序在 Client 模式下运行时，用户可以在客户端机器上直接与 Driver 程序通信，从而方便地进行调试和监视应用程序的执行情况。然而，由于 Driver 程序和 Executor 运行在不同的节点上，因此需要在网络之间传输数据和通信，这可能会增加网络带宽的消耗。

用户启动 Client 端，在 Client 端启动 Driver 进程。在 Driver 中启动或实例化 DAGScheduler 等组件，如图 9-3 所示。Spark 运行在 Client 的步骤如下。

1）Driver 在 Client 启动，做好准备工作，计划好任务的策略和方式（DAGScheduler）后向 Master

注册并申请运行 Executor 资源。

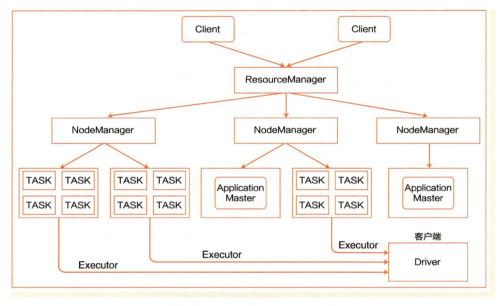

● 图 9-3 Spark 的 Client 运行模式

2）Worker 向 Master 注册，Master 通过指令让 Worker 启动 Executor。

3）Worker 收到指令后创建 ExecutorRunner 线程，进而 ExecutorRunner 线程启动 ExecutorBackend 进程。

4）ExecutorBackend 启动后，向 Client 端 Driver 进程内的 SchedulerBackend 注册，这样 Dirver 进程就可以发现计算资源了。

5）Driver 的 DAGScheduler 解析应用中的 RDD DAG 并生成相应的 Stage，每个 Stage 包含的 TaskSet 通过 TaskScheduler 分配给 Executor，在 Executor 内部启动线程池并行化执行 Task，同时 Driver 会密切注视，如果发现哪个 Execucutor 执行效率低，会分配其他 Execucutor 顶替执行，观察谁的效率更高（推测执行）。

6）计划中的所有 Stage 被执行完了之后，各个 Worker 汇报给 Driver，同时释放资源，Driver 确定都做完了，就向 Master 汇报。同时 Driver 在 Client 上，应用的执行进度 Client 也知道了。

9.2.4 Driver 运行在 Worker 节点

Spark 的运行模式中的 Driver 运行在 Worker 节点模式下，这意味着 Driver 程序和执行计算任务的 Executor 运行在同一个节点上。在 Worker 节点模式下，Driver 程序运行在集群中的一个工作节点上，而 Executor 则部署在同一个节点上。

在这种模式下，用户可以通过 Spark 的 API 将作业提交给集群管理器，并由集群管理器在合适的 Worker 节点上启动 Driver 程序。Driver 程序负责将任务分发到各个 Executor 上，并收集执行结果。它与 Spark 集群管理器和其他节点通过网络进行通信。

当 Spark 应用程序在 Worker 节点模式下运行时，由于 Driver 程序和 Executor 运行在同一节点上，因此可以减少网络带宽的消耗，提高了通信效率。此外，由于 Driver 程序运行在工作节点上，因此可以更方便地利用节点的计算资源，提高整体计算效率。

需要注意的是，由于 Driver 程序和 Executor 运行在同一节点上，因此当该节点发生故障或异常情况时，整个应用程序可能会受到影响。此外，如果应用程序需要处理大规模数据集，单个节点的计算和存储能力可能会成为瓶颈。

用户启动客户端，客户端提交应用程序给 Master，如图 9-4 所示。Spark 运行在 Worker 节点的步骤如下。

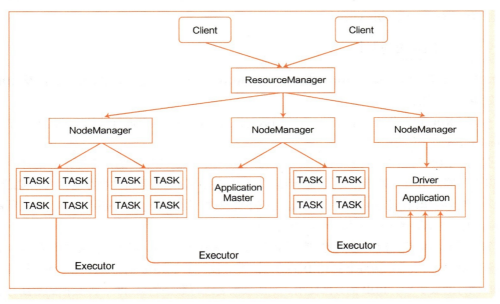

● 图 9-4　Spark 的 Driver 运行模式

1）Master 调度应用，指定一个 Worker 节点启动 Driver，即 Scheduler-Backend。

2）Worker 接收到 Master 命令后创建 DriverRunner 线程，在 DriverRunner 线程内创建 Scheduler-Backend 进程，Dirver 充当整个作业的主控进程。

3）Master 指定其他 Worker 节点启动 Executor，此处流程和上面相似，Worker 创建 ExecutorRunner 线程，启动 ExecutorBackend 进程。

4）ExecutorBackend 启动后，向 Client 端 Driver 进程内的 SchedulerBackend 注册，这样 Dirver 进程就可以发现计算资源了。

5）Driver 的 DAGScheduler 解析应用中的 RDD DAG 并生成相应的 Stage，每个 Stage 包含的 TaskSet 通过 TaskScheduler 分配给 Executor，在 Executor 内部启动线程池并行化执行 Task，同时 Driver 会密切注视，如果发现哪个 Executor 执行效率低，会分配其他 Executor 顶替执行，观察谁的效率更高（推测执行）。

6）计划中的所有 Stage 被执行完了之后，各个 Worker 汇报给 Driver，同时释放资源，Driver 确定都做完了，就向 Master 汇报。客户也会跳过 Master 直接和 Drive 通信了解任务的执行进度。

9.3 Spark 算子 RDD

RDD（Resilient Distributed Dataset）即弹性分布式数据集，是 Spark 中最基本的数据抽象，它代表一个不可变、可分区、里面的元素可并行计算的集合。RDD 具有数据流模型的特点：自动容错、位置感知性调度和可伸缩性。

对 RDD 进行改动，只能通过 RDD 的转换操作，由一个 RDD 得到一个新的 RDD，新的 RDD 包含了从其他 RDD 衍生所必需的信息。

9.3.1 RDD 的属性

在本节中，我们将深入探讨 RDD 的属性，包括 RDD 的创建、转化和动作等操作，以及 RDD 的分区、依赖和存储等属性。

1）分区：RDD 可以分为多个分区，每个分区可以在不同的节点上进行计算。分区的数量可以通过 getNumPartitions() 方法获取。

2）弹性：RDD 可以在内存和磁盘之间进行数据交换，以适应不同的计算需求。这种弹性使得 RDD 可以在不同的节点上进行计算，并且可以在计算过程中根据需要将数据存储在内存或磁盘中。

3）不可变性：RDD 是不可变的，一旦创建就不能修改。这种不可变性使得 RDD 可以进行高效的并行计算，因为不需要考虑数据的修改和同步问题。

4）宽依赖和窄依赖：RDD 之间的依赖关系分为宽依赖和窄依赖，宽依赖需要进行数据的 shuffle，而窄依赖不需要。窄依赖是指每个父 RDD 的分区最多只被一个子 RDD 的分区使用，而宽依赖是指每个父 RDD 的分区被多个子 RDD 的分区使用。宽依赖需要进行数据的 shuffle，即数据的重新分区和排序，这会带来额外的开销和性能损失。

9.3.2 RDD 的依赖关系

1. 依赖

RDD 和它依赖的父 RDD 的关系有两种不同的类型，即窄依赖和宽依赖，如图 9-5 所示。

1）窄依赖：指的是每一个父 RDD 的 Partition 最多被子 RDD 的一个 Partition 使用。

2）宽依赖：指的是同一个父 RDD 的 Partition 会被多个子 RDD 的 Partition 使用。

2. Spark 任务中的 Stage

DAG（Directed Acyclic Graph）即有向无环图，原始的 RDD 通过一系列的转换就形成了 DAG，根据 RDD 之间的依赖关系的不同将 DAG 划分成不同的 Stage。对于窄依赖，Partition 的转换处理在 Stage 中完成计算；对于宽依赖，由于有 shuffle 的存在，只能在 Parent RDD 处理完成后，才能开始接下来的计算，因此宽依赖是划分 Stage 的依据，如图 9-6 所示。

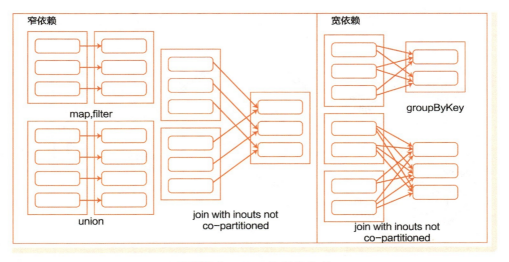

- 图 9-5　RDD 的宽窄依赖

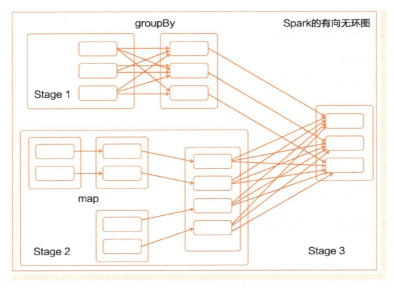

- 图 9-6　Spark 的有向无环图

9.3.3　RDD 的 shuffle 过程

Spark 是实现了 MapReduce 原语的一种通用实时计算框架。Spark 作业中 Map 阶段的 shuffle 称为 Shuffle Write，Reduce 阶段的 shuffle 称为 Shuffle Read，下面详细介绍 shuffle 的过程，如图 9-7 所示。

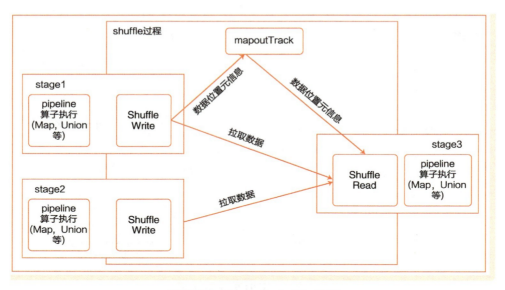

● 图 9-7　Spark 的 shuffle 过程

1. Shuffle Write

1）在分区之间重新分配数据并将文件写入磁盘。
2）每个 hash shuffle task 为每个 reduce task 创建一个文件。
3）sort shuffle task 创建一个文件，其中区域分配给 reducer。
4）sort shuffle 使用内存排序和溢出到磁盘，以获得最终结果。

2. Shuffle Read

1）获取文件并应用 reduce 逻辑。
2）如果需要数据有序，则对于任何类型的 shuffle，它在 reducer 侧排序。
3）在 Spark Sort Shuffle 是自版本 1.2 以来的默认值，但 Hash Shuffle 也可用。

9.3.4　RDD 的缓存和检查机制

1. RDD 的缓存机制

RDD 通过 cache 方法或者 persist 方法可以将前面的计算结果缓存，但并不是立即缓存，而是在接下来调用 action 类的算子的时候，该 RDD 将会被缓存在计算节点的内存中，并供后面使用。它既不是 transformation 也不是 action 类的算子。如图 9-8 所示。

2. RDD 的检查机制

检查点（本质是通过将 RDD 写入做检查点）是为了通过 Lineage（血统）做容错的辅助，Lineage 过长会造成容错成本过高，这样就不如在中间阶段做检查点容错，如果之后有节点出现问题而丢失分区，从做检查点的 RDD 开始重做 Lineage，就会减少开销。

第 9 章
数据处理分析引擎 Spark

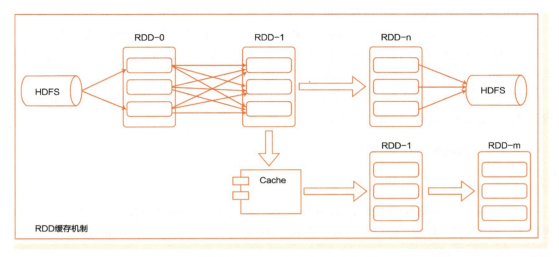

● 图 9-8　Spark 的 RDD 缓存机制

设置 checkpoint 的目录，可以是本地的文件夹，也可以是 HDFS。一般是在具有容错能力、高可靠的文件系统上（比如 HDFS、S3 等）设置一个检查点路径，用于保存检查点数据。

在设置检查点之后，该 RDD 之前的有依赖关系的父 RDD 都会被销毁，下次调用的时候直接从检查点开始计算。

9.4　Spark SQL

9.4.1　Spark SQL 概念

Spark SQL 是一种用于处理结构化数据的高性能分布式查询引擎，提供了一种强大的 SQL 查询功能和内置的数据处理功能。Spark SQL 还提供了 DataFrame 和 Dataset 这两个概念，它们是强类型的数据结构，类似于关系数据库中的表。DataFrame 和 Dataset 提供了更高级的 API，可以进行更复杂的数据操作和转换。

DataFrame 是一种以 RDD 为基础的分布式数据集，类似于传统数据库中的二维表格。DataFrame 与 RDD 的主要区别在于，DataFrame 带有 schema 元信息，即 DataFrame 所表示的二维表数据集的每一列都带有名称和类型。这使得 Spark SQL 得以洞察更多的结构信息，从而对藏于 DataFrame 背后的数据源以及作用于 DataFrame 之上的变换进行了针对性的优化，最终达到大幅提升运行时效率的目标。

Spark SQL 的特点如下。

1）高性能：Spark SQL 使用基于内存的计算模型，可以在内存中快速处理大规模数据集。此外，它还支持基于列存储的压缩格式，可以进一步提高查询性能。

2）多种数据源支持：Spark SQL 支持多种数据源，包括 Hive、JSON、Parquet、ORC、Avro、JDBC 等，可以方便地处理不同格式的数据。

3）SQL 查询支持：Spark SQL 支持标准的 SQL 查询语言，可以使用 SQL 语句进行数据查询和分析。此外，它还支持 DataFrame API 和 Dataset API，可以使用编程语言进行数据处理和分析。

4）集成性：Spark SQL 与 Spark 生态系统中的其他组件（如 Spark Streaming、MLlib、GraphX 等）集成紧密，可以方便地进行数据处理、机器学习和图计算等任务。

5）扩展性：Spark SQL 提供了丰富的 API 和插件机制，可以方便地扩展和定制功能。此外，它还支持自定义函数和聚合函数，可以满足各种数据处理需求。

9.4.2　Spark SQL 的架构

1. Spark SQL 的架构以及组件

1）SparkSession：SparkSession 是 Spark SQL 的入口点，它封装了 Spark SQL 的所有功能，并提供了一些方便的 API 来操作数据。SparkSession 负责创建 DataFrame、Dataset 和 SQLContext 等对象，以及管理 Spark 应用程序的生命周期。

2）DataFrame 和 Dataset：DataFrame 和 Dataset 是 Spark SQL 的核心数据结构，它们是一种分布式的、不可变的、面向列的数据集合。DataFrame 和 Dataset 可以从多种数据源中读取数据，并支持基于列的转换和操作。DataFrame 和 Dataset 还支持 SQL 查询和聚合操作，可以使用标准的 SQL 语句进行数据查询和分析。

3）SQLContext：SQLContext 是 Spark SQL 的上下文对象，它负责管理 Spark SQL 的元数据和执行 SQL 查询。SQLContext 可以从多种数据源中读取数据，并将其转换为 DataFrame 或 Dataset 对象。SQLContext 还支持注册 UDF 和 UDAF 等自定义函数，以满足各种数据处理需求。

4）Catalyst Optimizer：Catalyst Optimizer 是 Spark SQL 的查询优化器，它负责将 SQL 查询转换为物理执行计划，并对执行计划进行优化。Catalyst Optimizer 使用基于规则和代价的优化技术，可以自动推断查询的最优执行计划，并生成高效的代码。

5）DataSource API：DataSource API 是 Spark SQL 的数据源 API，它提供了一种统一的方式来读取和写入不同格式的数据。DataSource API 支持多种数据源，包括 Hive、JSON、Parquet、ORC、Avro、JDBC 等，可以方便地处理不同格式的数据。

6）Spark Core：Spark Core 是 Spark SQL 的底层引擎，它提供了分布式计算和数据处理的基础设施。Spark Core 使用基于内存的计算模型，可以在内存中快速处理大规模数据集。Spark Core 还支持任务调度、容错和数据分区等功能，可以保证 Spark 应用程序的高可用性和高性能。

2. Spark SQL 的流程

通过优化查询和并行计算，Spark SQL 能够有效地处理大规模的数据集，提供快速而可靠的数据查询和分析功能。Spark SQL 的运行流程可以大致分为以下几个步骤。

1）数据源加载：Spark SQL 会从指定的数据源中加载数据。这些数据源可以是文件系统、Hive 表、HBase 等。Spark SQL 支持多种数据格式，如 CSV、JSON、Parquet 等。

2）Schema 推断：在加载数据之后，Spark SQL 会尝试自动推断数据的结构，并为每个字段分配相应的数据类型。如果数据源中已经包含了模式信息，那么这一步可以被省略。

3）查询优化：一旦数据加载和模式推断完成，Spark SQL 会根据用户提供的 SQL 语句进行查询优化。这个过程包括查询重写、谓词下推、投影消除等操作，旨在提高查询性能。

4）执行计划生成：在查询优化完成后，Spark SQL 会生成一个执行计划。执行计划是一个逻辑执行的计划图，描述了如何对数据进行操作，以满足用户的查询需求。

5）分布式执行：一旦执行计划生成，Spark SQL 会将任务分发到集群中的不同节点上进行并行计算。每个节点会负责处理一部分数据，并生成局部的结果。

6）数据汇总和聚合：在各个节点上完成计算后，Spark SQL 会将结果数据汇总到驱动程序（Driver）节点上，并进行进一步的聚合计算（如果有需要的话）。

7）结果返回：最后，Spark SQL 将计算的结果返回给用户，并根据需要进行格式转换和数据输出。

这些步骤展示了 Spark SQL 的一般运行流程。通过优化查询和并行计算，Spark SQL 能够有效地处理大规模的数据集，提供快速而可靠的数据查询和分析功能。

▶▶ 9.4.3 DataSets 和 DataFrames

DataFrame 是一种以 RDD 为基础的分布式数据集，类似于传统数据库的二维表格，DataFrame 带有 Schema 元信息，即 DataFrame 所表示的二维表数据集的每一列都带有名称和类型。DataFrame 可以从很多数据源构建，比如已经存在的 RDD、结构化文件、外部数据库、Hive 表。DataFrame 与 RDD 的区别如图 9-9 所示。

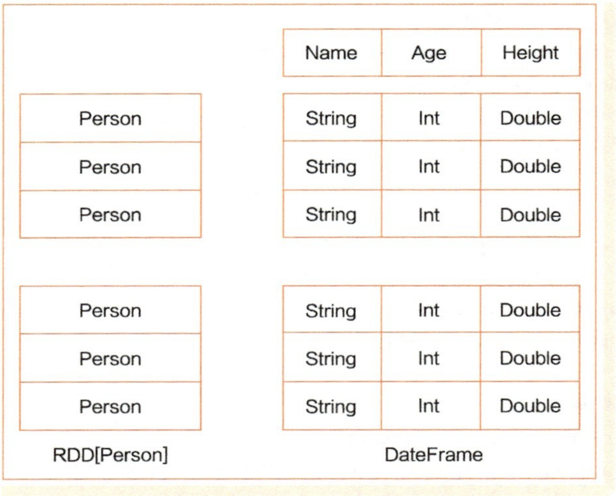

Person		Name	Age	Height
Person		String	Int	Double
Person		String	Int	Double
Person		String	Int	Double
Person		String	Int	Double
Person		String	Int	Double
Person		String	Int	Double
RDD[Person]			DateFrame	

● 图 9-9　DataFrame 与 RDD 的区别

从图 9-9 可以看出 DataFrame 与 RDD 的区别如下。

1）RDD 可看作是分布式的对象集合，Spark 并不知道对象的详细模式信息，DataFrame 可看作是分布式的 Row 对象的集合，其提供了由列组成的详细模式信息，使得 Spark SQL 可以进行某些形式的执行优化。DataFrame 和普通的 RDD 逻辑框架的区别如下所示。

2）左侧的 RDD [Person] 虽然以 Person 为类型参数，但 Spark 框架本身不了解 Person 类的内部结构。而右侧的 DataFrame 却提供了详细的结构信息，使得 Spark SQL 可以清楚地知道该数据集中包含哪些列，每列的名称和类型各是什么。DataFrame 多了数据的结构信息，即 schema。这样看起来就像一张表了，DataFrame 还配套了新的操作数据的方法，DataFrame API。

3）RDD 是分布式的 Java 对象的集合。DataFrame 是分布式的 Row 对象的集合。DataFrame 除了提供了比 RDD 更丰富的算子以外，更重要的特点是提升执行效率、减少数据读取以及执行计划的优化。

4）有了 DataFrame 这个数据抽象后，可以用 SQL 来处理数据了，对开发者来说，易用性有了很大的提升。不仅如此，通过 DataFrame API 或 SQL 处理数据，会自动经过 Spark 优化器（Catalyst）的优化，即使你写的程序或 SQL 不高效，也可以运行得很快。

Dataset 是数据的分布式集合。Dataset 是在 Spark 1.6 中添加的一个新接口，是 DataFrame 之上更高一级的抽象。它提供了 RDD 的优点以及 Spark SQL 优化后的执行引擎的优点。一个 DataSet 可以从 JVM 对象构造，然后使用函数转换去操作。Dataset API 支持 Scala 和 Java。

9.4.4 Spark SQL 示例

```java
import org.apache.spark.sql.Dataset;
import org.apache.spark.sql.Row;
import org.apache.spark.sql.SparkSession;

public class SparkSQLExample {
    public static void main(String[] args) {
        // 创建 SparkSession 对象
        SparkSession spark = SparkSession.builder()
            .appName("SparkSQLExample")
            .master("local[*]")
            .getOrCreate();

        // 读取数据
        Dataset<Row> df = spark.read()
            .format("csv")
            .option("header", "true")
            .load("path/to/file.csv");

        // 执行 SQL 查询
        df.createOrReplaceTempView("people");
        Dataset<Row> result = spark.sql("SELECT * FROM people WHERE age > 18");

        // 输出结果
        result.show();

        // 关闭 SparkSession 对象
        spark.stop();
    }
}
```

以上代码使用 SparkSession 对象读取 CSV 文件，并执行 SQL 查询。首先，使用 SparkSession.builder() 方法创建一个 SparkSession 对象，并指定应用程序名称和运行模式。然后，使用 spark.read() 方法读取 CSV 文件，并将其加载到 DataFrame 对象中。接下来，使用 df.createOrReplaceTempView() 方法将 DataFrame 对象注册为一个临时表，以便我们可以在 SQL 查询中使用它。最后，使用 spark.sql() 方法执行 SQL 查询，并使用 result.show() 方法输出结果。

9.5 Spark Streaming

9.5.1 Spark Streaming 介绍

Spark Streaming 是核心 Spark API 的扩展，可实现可扩展、高吞吐量、可容错的实时数据流处理。数据可以从 Kafka、Flume、TCP 套接字等众多来源获取，并且可以使用由高级函数开发的复杂算法进行流数据处理。最后，处理后的数据可以被推送到文件系统、数据库和实时仪表板，如图 9-10 所示。

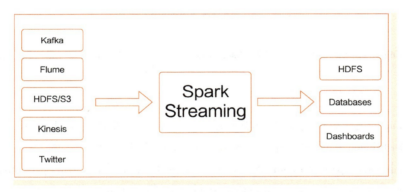

● 图 9-10　Spark Streaming 示意图

1. Spark Streaming 的特点

在本节中，我们将深入探讨 Spark Streaming 的特点，包括其核心概念、架构和数据流处理方式。你将了解如何使用 Spark Streaming 处理实时数据流，并了解其内部的消息处理机制。Spark Streaming 的特点如下。

1）高吞吐量和低延迟：Spark Streaming 可以处理高吞吐量的数据流，并且具有低延迟的特点。这使得 Spark Streaming 非常适合处理实时数据流，例如网络日志、传感器数据等。

2）容错性：Spark Streaming 具有高度的容错性，可以自动恢复失败的节点和任务。这种容错性使得 Spark Streaming 可以处理大规模的数据流，并且可以保证数据的完整性和准确性。

3）灵活性：Spark Streaming 可以与其他 Spark 组件集成，例如 Spark SQL、Spark MLlib 等。这种灵活性使得 Spark Streaming 可以处理各种类型的数据流，并且可以进行复杂的数据处理和分析。

4）可扩展性：Spark Streaming 可以水平扩展，也可以通过增加节点和资源来处理更大规模的数据流。这种可扩展性使得 Spark Streaming 可以应对不断增长的数据流,并且可以保持高吞吐量和低延迟的特点。

以上是 Spark Streaming 的特点和作用的简单说明。Spark Streaming 可以通过各种操作进行数据处理和转换,例如 map()、filter()、reduceByKey()等。这些操作可以在 DStream 上进行链式调用,形成一个操作流程。DStream 的操作也是惰性的,只有在需要计算结果时才会进行实际的计算。这种惰性计算使得 Spark Streaming 可以进行高效的优化和调度,以提高计算性能。

2. Spark Streaming 的内部结构

在内部,它的工作原理如下。Spark Streaming 接收实时输入数据流,并将数据切分成批,然后由 Spark 引擎对其进行处理,最后生成"批"形式的结果流,如图 9-11 所示。

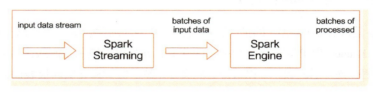

● 图 9-11　Spark Streaming 结构

Spark Streaming 将连续的数据流抽象为 Discretizedstream 或 DStream。在内部,DStream 由一个 RDD 序列表示。

3. Spark Streaming 工作原理

Spark Streaming 的工作原理是将实时数据流划分为一系列小批次(batch),每个批次都是一个 RDD(Resilient Distributed Dataset)。Spark Streaming 将这些小批次 RDD 作为输入,通过 Spark 引擎进行处理,生成实时结果。Spark Streaming 的核心是 StreamingContext 对象,它是 Spark Streaming 应用程序的入口点。StreamingContext 对象可以从 SparkContext 对象中创建,它定义了批处理间隔(batch interval)和数据源(data source)等参数。

Spark Streaming 应用程序通过 StreamingContext 对象创建 DStream 对象,DStream 对象是对实时数据流的抽象,它可以从数据源中读取数据,并进行转换和操作。DStream 对象支持基于时间窗口和滑动窗口的转换和操作,可以对数据流进行实时的聚合、过滤、转换等操作。最后,Spark Streaming 应用程序通过 DStream 对象生成实时结果,可以将结果输出到控制台、文件、数据库等。Spark Streaming 应用程序启动后,会不断地读取数据源中的数据,生成实时结果,直到应用程序被停止。

4. Spark Streaming 离散流 DStream

StreamingContext 会根据设置的批处理的时间间隔将产生的 RDD 归为一批,这一批 RDD 就是一个 DStream,DStream 可以通过算子操作转化为另一个 DStream,DStream 是 Spark Streaming 对流式数据的基本抽象。它表示连续的数据流,这些连续的数据流可以是从数据源接收的输入数据流,也可以是通过对输入数据流执行转换操作而生成的经处理的数据流。在内部,DStream 由一系列连续的 RDD 表示,如图 9-12 所示。

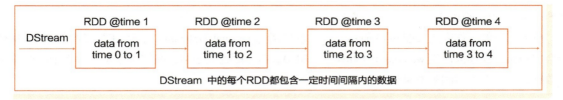

● 图 9-12　Spark Streaming 离散流 DStream

9.5.2　DStream 转换操作

下面是一个示例代码，它使用 socketTextStream() 方法创建了一个 DStream，从 TCP Socket 读取数据，并使用 map() 操作将输入数据中的每个元素转换为大写字母。

```java
import org.apache.spark.SparkConf;
import org.apache.spark.streaming.Durations;
import org.apache.spark.streaming.api.java.JavaDStream;
import org.apache.spark.streaming.api.java.JavaStreamingContext;

public class SparkStreamingExample {
    public static void main(String[] args) throws InterruptedException {
        // 创建 Spark Streaming 上下文
        SparkConf conf = new SparkConf().setAppName("SparkStreamingExample");
        JavaStreamingContext jssc = new JavaStreamingContext(conf, Durations.seconds(1));

        // 创建 DStream,从 TCP Socket 读取数据
        JavaDStream<String> lines = jssc.socketTextStream("localhost", 9999);

        // 对每个元素执行 map 操作,将其转换为大写字母
        JavaDStream<String> upperCaseLines = lines.map(String::toUpperCase);

        // 输出转换后的数据
        upperCaseLines.print();

        // 启动 Spark Streaming 应用程序
        jssc.start();
        jssc.awaitTermination();
    }
}
```

案例中首先创建了一个 Spark Streaming 上下文，并使用 socketTextStream() 方法创建了一个 DStream，从 TCP Socket 读取数据。然后，对每个元素执行 map() 操作，将其转换为大写字母，并使用 print() 方法输出转换后的数据。最后，启动 Spark Streaming 应用程序并等待其终止。

9.5.3　Spark Streaming 窗口操作

Spark Streaming 还提供了窗口计算功能，允许设置窗口的大小和滑动窗口的间隔来动态地获取当

前 Steaming 的允许状态。

滑动窗口的工作方式如图 9-13 所示，每当窗口滑过 original DStream 时，落在窗口内的源 RDD 被组合并被执行操作以产生 windowed DStream 的 RDD。在上面的例子中，操作应用于最近 3 个时间单位的数据，并以 2 个时间单位滑动。

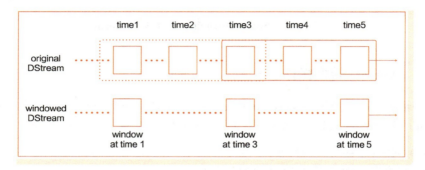

● 图 9-13　滑动窗口的工作方式

下面是一个示例代码，它使用 window() 方法创建了一个窗口，对窗口内的数据进行操作，并使用 reduceByKey() 操作计算每个单词出现的次数。

```java
import org.apache.spark.SparkConf;
import org.apache.spark.streaming.Durations;
import org.apache.spark.streaming.api.java.JavaDStream;
import org.apache.spark.streaming.api.java.JavaPairDStream;
import org.apache.spark.streaming.api.java.JavaStreamingContext;
import scala.Tuple2;

public class SparkStreamingExample {
    public static void main(String[] args) throws InterruptedException {
        // 创建 Spark Streaming 上下文
        SparkConf conf = new SparkConf().setAppName("SparkStreamingExample");
        JavaStreamingContext jssc = new JavaStreamingContext(conf, Durations.seconds(1));

        // 创建 DStream,从 TCP Socket 读取数据
        JavaDStream<String> lines = jssc.socketTextStream("localhost", 9999);

        // 将每行数据拆分为单词
        JavaDStream<String> words = lines.flatMap(line -> Arrays.asList(line.split(" ")).iterator());

        // 创建窗口,对窗口内的数据进行操作
        JavaDStream<String> windowedWords = words.window(Durations.seconds(10), Durations.seconds(5));
```

```
        // 计算每个单词出现的次数
        JavaPairDStream<String, Integer> wordCounts = windowedWords.mapToPair(word -> new
Tuple2<>(word, 1))
                .reduceByKey((count1, count2) -> count1 + count2);

        // 输出计算结果
        wordCounts.print();

        // 启动 Spark Streaming 应用程序
        jssc.start();
        jssc.awaitTermination();
    }
}
```

案例中首先创建了一个 Spark Streaming 上下文,并使用 socketTextStream( ) 方法创建了一个 DStream,从 TCP Socket 读取数据。然后,将每行数据拆分为单词,并使用 window( ) 方法创建了一个窗口,对窗口内的数据进行操作。接着,使用 mapToPair( ) 操作将每个单词映射为一个键值对,其中键为单词本身,值为 1,然后使用 reduceByKey( ) 操作计算每个单词出现的次数。最后,使用 print( ) 方法输出计算结果。

▶▶ 9.5.4　DStream 输入

DStream 可以从各种数据源中创建,例如 Kafka、Flume、HDFS、socket 等。它代表从服务器获取的数据流。每一个输入流 DStream 和一个 Receiver 对象相关联,这个 Receiver 从源中获取数据,并将数据存入内存中用于处理。

DStream 输入案例

1)文件流:使用 Spark Streaming 监控文件作为数据流,您需要使用 JavaStreamingContext 的 textFileStream( ) 方法。该方法接受一个字符串参数,表示监控的目录路径。下面是一个使 Spark Streaming 监控文件作为数据流的示例代码。

```
import org.apache.spark.SparkConf;
import org.apache.spark.streaming.Durations;
import org.apache.spark.streaming.api.java.JavaStreamingContext;
import org.apache.spark.streaming.api.java.JavaDStream;
import org.apache.spark.streaming.api.java.JavaRDD;
import org.apache.spark.streaming.api.java.function.Function;

public class SparkStreamingFileMonitorExample {
    public static void main(String[] args) throws InterruptedException {
        // 创建 Spark Streaming 上下文
        SparkConf conf = new SparkConf().setAppName("SparkStreamingFileMonitorExample");
        JavaStreamingContext jssc = new JavaStreamingContext(conf, Durations.seconds(1));
```

```java
        // 设置检查点目录
        jssc.checkpoint("/path/to/checkpoint");

        // 监控文件作为数据流
        JavaDStream<String> lines = jssc.textFileStream("/path/to/monitor");

        // 对数据流进行处理
        JavaDStream<String> words = lines.flatMap(line -> Arrays.asList(line.split(" ")).iterator());
        JavaDStream<String> filteredWords = words.filter(word -> word.startsWith("a"));

        // 输出处理结果
        filteredWords.print();

        // 启动 Spark Streaming 应用程序
        jssc.start();
        jssc.awaitTermination();
    }
}
```

在这个示例代码中,我们首先创建了一个 Spark Streaming 上下文,并设置了应用程序名称和批处理间隔。然后,使用 checkpoint( )方法设置了检查点目录的路径。接着,使用 textFileStream( )方法监控了一个目录,并将其作为数据流。然后,对数据流进行了处理,包括将每行拆分为单词、过滤以字母 a 开头的单词等。最后,输出了处理结果。需要注意的是,需要将/path/to/monitor 替换为要监控的目录路径。

2)RDD 队列流:使用 Spark Streaming 队列流,需要使用 JavaStreamingContext 的 queueStream( )方法。该方法接受一个 Queue<JavaRDD>类型的参数,其中 T 是要处理的数据类型。下面是一个 Spark Streaming 队列流的示例代码。

```java
import java.util.Arrays;
import java.util.LinkedList;
import java.util.Queue;
import org.apache.spark.SparkConf;
import org.apache.spark.streaming.Durations;
import org.apache.spark.streaming.api.java.JavaStreamingContext;
import org.apache.spark.streaming.api.java.JavaDStream;
import org.apache.spark.streaming.api.java.JavaRDD;
import org.apache.spark.streaming.api.java.function.Function;

public class SparkStreamingQueueExample {
    public static void main(String[] args) throws InterruptedException {
        // 创建 Spark Streaming 上下文
        SparkConf conf = new SparkConf().setAppName("SparkStreamingQueueExample");
        JavaStreamingContext jssc = new JavaStreamingContext(conf, Durations.seconds(1));
```

```java
        // 设置检查点目录
        jssc.checkpoint("/path/to/checkpoint");

        // 创建一个包含 RDD 的队列
        Queue<JavaRDD<String>> queue = new LinkedList<JavaRDD<String>>();
        queue.add(jssc.sparkContext().parallelize(Arrays.asList("hello world")));
        queue.add(jssc.sparkContext().parallelize(Arrays.asList("goodbye world")));

        // 创建队列流
        JavaDStream<String> lines = jssc.queueStream(queue);

        // 对数据流进行处理
        JavaDStream<String> words = lines.flatMap(line -> Arrays.asList(line.split(" ")).iterator());
        JavaDStream<String> filteredWords = words.filter(word -> word.startsWith("h"));

        // 输出处理结果
        filteredWords.print();

        // 启动 Spark Streaming 应用程序
        jssc.start();
        jssc.awaitTermination();
    }
}
```

示例中首先创建了一个 Spark Streaming 上下文,并设置了应用程序名称和批处理间隔。然后,使用 checkpoint() 方法设置了检查点目录的路径。接着,创建了一个包含两个 RDD 的队列,并使用 queueStream() 方法创建了一个队列流。然后,对数据流进行了处理,包括将每行拆分为单词、过滤以字母 h 开头的单词等。最后,输出了处理结果。

3) 套接字流:使用 Spark Streaming 队列流,需要使用 JavaStreamingContext 的 socketTextStream() 方法创建一个套接字流,通过监听 Socket 端口来接收数据,使用 nc 工具给指定端口发送数据。下面是一个 Spark Streaming 套接字流的示例代码。

```java
import org.apache.spark.SparkConf;
import org.apache.spark.streaming.Durations;
import org.apache.spark.streaming.api.java.JavaStreamingContext;
import org.apache.spark.streaming.api.java.JavaDStream;

public class SparkStreamingSocketExample {
    public static void main(String[] args) throws InterruptedException {
        // 创建 Spark Streaming 上下文
        SparkConf conf = new SparkConf().setAppName("SparkStreamingSocketExample");
        JavaStreamingContext jssc = new JavaStreamingContext(conf, Durations.seconds(1));

        // 设置检查点目录
```

```
        jssc.checkpoint("/path/to/checkpoint");

        // 创建套接字流
        JavaDStream<String> lines = jssc.socketTextStream("localhost", 9999);

        // 对数据流进行处理
        JavaDStream<String> words = lines.flatMap(line -> Arrays.asList(line.split(" ")).
iterator());
        JavaDStream<String> filteredWords = words.filter(word -> word.startsWith("a"));

        // 输出处理结果
        filteredWords.print();

        // 启动 Spark Streaming 应用程序
        jssc.start();
        jssc.awaitTermination();
    }
}
```

示例中首先创建了一个 Spark Streaming 上下文,并设置了应用程序名称和批处理间隔。然后,使用 checkpoint( )方法设置了检查点目录的路径。接着,使用 socketTextStream( )方法创建了一个套接字流,并指定了主机名和端口号。然后,对数据流进行了处理,包括将每行拆分为单词、过滤以字母 a 开头的单词等。最后,输出了处理结果。需要注意的是,需要将 localhost 和 9999 替换为要连接的主机名和端口号,并在 TODO 注释下方添加自己的 Spark Streaming 应用程序代码。

9.5.5 DStream 输出

DStream 是 Spark Streaming 中最基本的抽象,它代表了一个连续的数据流。在创建 DStream 之后,可以对其进行各种转换操作,例如 map、filter、reduceByKey 等,以便对数据流进行处理。最后,可以使用输出操作,例如 print、saveAsTextFiles 等,将处理结果输出到控制台或文件中。

下面是一个 DStream 将数据写入 MySQL 数据库的示例代码。

```
import java.sql.Connection;
import java.sql.DriverManager;
import java.sql.PreparedStatement;
import java.sql.SQLException;
import java.util.Arrays;

import org.apache.spark.SparkConf;
import org.apache.spark.streaming.Durations;
import org.apache.spark.streaming.api.java.JavaStreamingContext;
import org.apache.spark.streaming.api.java.JavaDStream;

public class SparkStreamingDStreamOutputMySQLExample {
```

```java
public static void main(String[] args) throws InterruptedException {
    // 创建 Spark Streaming 上下文
    SparkConf conf = new SparkConf().setAppName("SparkStreamingDStreamOutputMySQLExample");
    JavaStreamingContext jssc = new JavaStreamingContext(conf, Durations.seconds(1));

    // 设置检查点目录
    jssc.checkpoint("/path/to/checkpoint");

    // 创建套接字流
    JavaDStream<String> lines = jssc.socketTextStream("localhost", 9999);

    // 对数据流进行处理
    JavaDStream<String> words = lines.flatMap(line -> Arrays.asList(line.split(" ")).iterator());
    JavaDStream<String> filteredWords = words.filter(word -> word.startsWith("a"));

    // 输出处理结果到 MySQL
    filteredWords.foreachRDD(rdd -> {
        rdd.foreachPartition(partition -> {
            Connection conn = DriverManager.getConnection("jdbc:mysql://localhost:3306/mydatabase", "myuser", "mypassword");
            PreparedStatement stmt = conn.prepareStatement("INSERT INTO mytable (word) VALUES (?)");
            while (partition.hasNext()) {
                String word = partition.next();
                stmt.setString(1, word);
                stmt.executeUpdate();
            }
            stmt.close();
            conn.close();
        });
    });

    // 启动 Spark Streaming 应用程序
    jssc.start();
    jssc.awaitTermination();
}
```

示例中首先创建了一个 Spark Streaming 上下文，并设置了应用程序名称和批处理间隔。然后，使用 checkpoint( ) 方法设置了检查点目录的路径。接着，使用 socketTextStream( ) 方法创建了一个套接字流，并指定了主机名和端口号。然后，对数据流进行了处理，包括将每行拆分为单词、过滤以字母 a 开头的单词等。最后，使用 foreachRDD( ) 方法将处理结果输出到 MySQL 中。需要注意的是，需要将 localhost 和 9999 替换为要连接的主机名和端口号，并将 jdbc:mysql://localhost:3306/mydatabase、myuser 和 mypassword 替换为自己的 MySQL 数据库连接信息和凭据。

9.5.6 DSFrame 和 SQL 操作

使用 DataFrames 和 SQL 操作来处理流数据。必须使用当前的 StreamingContext 对应的 SparkContext 创建一个 SparkSession。此外，必须这样做的另一个原因是使得应用可以在 Driver 程序故障时得以重新启动，这是通过创建一个可以延迟实例化的单例 SparkSession 来实现的。

下面是一个 DataFrames 和 SQL 示例代码。

```java
import org.apache.spark.sql.Dataset;
import org.apache.spark.sql.Row;
import org.apache.spark.sql.SparkSession;

public class SparkDataFrameSQLExample {
    public static void main(String[] args) {
        // 创建 SparkSession
        SparkSession spark = SparkSession.builder()
            .appName("SparkDataFrameSQLExample")
            .getOrCreate();

        // 读取 CSV 文件并创建 DataFrame
        Dataset<Row> df = spark.read()
            .option("header", "true")
            .option("inferSchema", "true")
            .csv("/path/to/csv");

        // 注册 DataFrame 为临时表
        df.createOrReplaceTempView("mytable");

        // 执行 SQL 查询
        Dataset<Row> result = spark.sql("SELECT COUNT(*) FROM mytable WHERE age > 30");

        // 输出查询结果
        result.show();

        // 关闭 SparkSession
        spark.stop();
    }
}
```

示例中首先创建了一个 SparkSession，并设置了应用程序名称。然后，使用 read() 方法读取了一个 CSV 文件，并创建了一个 DataFrame。使用 createOrReplaceTempView() 方法将 DataFrame 注册为临时表。然后，使用 sql() 方法执行了一个 SQL 查询，并将结果保存在一个新的 DataFrame 中。最后，使用 show() 方法输出了查询结果。需要注意的是，需要将/path/to/csv 替换为要读取的 CSV 文件的路径，并在 SQL 查询中使用自己的表名和列名。

9.5.7 Spark Streaming 检查点

为了保证 Spark Streaming 应用程序的容错性和可靠性，Spark 提供了检查点机制。检查点是将应用程序的状态保存到可靠的存储介质中，以便在应用程序失败时能够恢复状态并继续处理数据。在 Spark Streaming 中，检查点用于保存应用程序的元数据和状态信息，例如 DStream 操作的元数据、窗口操作的状态等。检查点可以在应用程序启动时设置，也可以在运行时动态设置。在示例代码中，我们使用了 checkpoint( ) 方法设置了检查点目录的路径，以便在应用程序失败时能够恢复状态。需要注意的是，检查点目录应该位于可靠的存储介质上，例如 HDFS 或 S3，以确保数据的可靠性和持久性。

下面是 Spark Streaming 设置检查点的示例代码。

```java
import org.apache.spark.SparkConf;
import org.apache.spark.streaming.Durations;
import org.apache.spark.streaming.api.java.JavaStreamingContext;

public class SparkStreamingCheckpointExample {
    public static void main(String[] args) throws InterruptedException {
        // 创建 Spark Streaming 上下文
        SparkConf conf = new SparkConf().setAppName("SparkStreamingCheckpointExample");
        JavaStreamingContext jssc = new JavaStreamingContext(conf, Durations.seconds(1));

        // 设置检查点目录
        jssc.checkpoint("/path/to/checkpoint");

        // TODO: 在这里添加自己的 Spark Streaming 应用程序代码

        // 启动 Spark Streaming 应用程序
        jssc.start();
        jssc.awaitTermination();
    }
}
```

示例中首先创建了一个 Spark Streaming 上下文，并设置了应用程序名称和批处理间隔。然后，使用 checkpoint( ) 方法设置了检查点目录的路径。最后，启动了 Spark Streaming 应用程序。需要注意的是，需要在 TODO 注释下方添加自己的 Spark Streaming 应用程序代码。

9.6 Spark Streaming 接收 Flume 数据实战

使用 Spark Streaming 接收 Flume 数据，需要使用 Spark Streaming 的 Flume 集成 API。Spark Streaming 的 Flume 集成 API 提供了两种方式来接收 Flume 数据：使用 Flume 的 Avro Source 和使用 Spark Streaming 的 Polling Stream。

1. 使用 flume pull 方式

设置 Flume 的配置文件 flume_pull_streaming.conf，以下是一个 Flume 配置文件，用于 Spark 将数据

拉取到 Spark Streaming，示例代码如下。

```
simple-agent.sources = netcat-source
simple-agent.sinks = spark-sink
simple-agent.channels = memory-channel

simple-agent.sources.netcat-source.type = netcat
simple-agent.sources.netcat-source.bind = bigdata.ibeifeng.com
simple-agent.sources.netcat-source.port = 44444

simple-agent.sinks.spark-sink.type = org.apache.spark.streaming.flume.sink.SparkSink
simple-agent.sinks.spark-sink.hostname = bigdata.ibeifeng.com
simple-agent.sinks.spark-sink.port = 41414

simple-agent.channels.memory-channel.type = memory

simple-agent.sources.netcat-source.channels = memory-channel
```

使用 Java 语言编写 Spark Streaming 接收 Flume pull 方式的数据，示例代码如下。

```java
import org.apache.spark.SparkConf;
import org.apache.spark.streaming.Duration;
import org.apache.spark.streaming.api.java.JavaDStream;
import org.apache.spark.streaming.api.java.JavaStreamingContext;
import org.apache.spark.streaming.flume.FlumeUtils;
import org.apache.spark.streaming.flume.SparkFlumeEvent;

public class SparkStreamingFlumePull {
    public static void main(String[] args) {
        // 创建一个 SparkConf 对象
        SparkConf conf = new SparkConf().setAppName("SparkStreamingFlumePull").setMaster("local[2]");
        // 创建一个 JavaStreamingContext 对象
        JavaStreamingContext jssc = new JavaStreamingContext(conf, new Duration(1000));
        // 创建一个 Flume pull 流
        JavaDStream<SparkFlumeEvent> flumeStream = FlumeUtils.createPollingStream(jssc, "localhost", 41414);
        // 输出 Flume 事件
        flumeStream.print();
        // 启动 Streaming 应用程序
        jssc.start();
        try {
            jssc.awaitTermination();
        } catch (InterruptedException e) {
            e.printStackTrace();
        }
    }
}
```

在这段代码中，我们首先创建了一个 SparkConf 对象，并设置了应用程序的名称和运行模式。然后，创建了一个 JavaStreamingContext 对象，并设置了批处理间隔为 1 秒。接着，使用 FlumeUtils 类的 createStream 方法创建了一个 Flume pull 流，指定了 Flume 的主机名和端口号。然后，使用 JavaDStream 对象的 print 方法输出 Flume 事件。最后，启动了 Streaming 应用程序，并等待其终止。

2. flume-push 方式

设置 Flume 的配置文件 flume_push_streaming.conf，以下是一个 Flume 配置文件，用于将数据推送到 Spark Streaming，示例代码如下。

```
simple-agent.sources = netcat-source
simple-agent.sinks = avro-sink
simple-agent.channels = memory-channel

simple-agent.sources.netcat-source.type = netcat
simple-agent.sources.netcat-source.bind = bigdata.ibeifeng.com
simple-agent.sources.netcat-source.port = 44444

simple-agent.sinks.avro-sink.type = avro
simple-agent.sinks.avro-sink.hostname = 192.168.81.1
simple-agent.sinks.avro-sink.port = 41414

simple-agent.channels.memory-channel.type = memory

simple-agent.sources.netcat-source.channels = memory-channel
```

在这个配置文件中，首先定义了三个组件：avro-source、spark-sink 和 memory-channel。然后，配置了 avro-source 组件，指定了其类型为 avro，并设置了其绑定的主机名和端口号。接着，配置了 spark-sink 组件，指定了其类型为 SparkSink，并设置了其绑定的主机名和端口号。最后，配置了 memory-channel 组件，指定了其类型为 memory，并设置了其缓冲区大小和事务容量。将 avro-source 和 spark-sink 组件绑定到了 memory-channel 组件上。

使用 Java 语言编写 Spark Streaming 接收 flume-push 方式的数据，示例代码如下。

```java
import org.apache.spark.SparkConf;
import org.apache.spark.streaming.Duration;
import org.apache.spark.streaming.api.java.JavaDStream;
import org.apache.spark.streaming.api.java.JavaStreamingContext;
import org.apache.spark.streaming.flume.FlumeUtils;
import org.apache.spark.streaming.flume.SparkFlumeEvent;

public class SparkStreamingFlumePush {
    public static void main(String[] args) {
        // 创建一个 SparkConf 对象
        SparkConf conf = new SparkConf().setAppName("SparkStreamingFlumePush").setMaster("local[2]");
```

```java
        // 创建一个 JavaStreamingContext 对象
        JavaStreamingContext jssc = new JavaStreamingContext(conf, new Duration(1000));
        // 创建一个 Flume push 流
        JavaDStream<SparkFlumeEvent> flumeStream = FlumeUtils.createStream(jssc, "local-host", 41414);
        // 输出 Flume 事件
        flumeStream.print();
        // 启动 Streaming 应用程序
        jssc.start();
        try {
            jssc.awaitTermination();
        } catch (InterruptedException e) {
            e.printStackTrace();
        }
    }
}
```

在这段代码中,首先创建了一个 SparkConf 对象,并设置了应用程序的名称和运行模式。然后,创建了一个 JavaStreamingContext 对象,并设置了批处理间隔为 1 秒。接着,使用 FlumeUtils 类的 createStream 方法创建了一个 Flume push 流,指定了 Flume 的主机名和端口号。然后,使用 JavaDStream 对象的 print 方法输出 Flume 事件。最后,启动了 Streaming 应用程序,并等待其终止。

9.7 Spark Streaming 接收 Kafka 数据实战

在这个实战中,我们将介绍如何使用 Spark Streaming 接收来自 Kafka 的数据,并进行实时处理。通过本实战,读者将了解如何将 Spark Streaming 与 Kafka 集成,实现数据的高效传输和处理。

Spark Streaming 接收 Kafka 数据,需要使用 Spark Streaming 的 Kafka 集成 API。Spark Streaming 的 Kafka 集成 API 提供了两种方式来接收 Kafka 数据:直接使用 Kafka 的高级 API 和使用 Spark Streaming 的 Direct API。

1. receiver 方式

使用 Java 语言编写 Spark Streaming,使用 receiver 方式接收 Kafka 数据,示例代码如下。

```java
import java.util.HashMap;
import java.util.Map;
import org.apache.kafka.common.serialization.StringDeserializer;
import org.apache.spark.SparkConf;
import org.apache.spark.streaming.Durations;
import org.apache.spark.streaming.api.java.JavaInputDStream;
import org.apache.spark.streaming.api.java.JavaPairDStream;
import org.apache.spark.streaming.api.java.JavaStreamingContext;
import org.apache.spark.streaming.kafka010.ConsumerStrategies;
```

```java
import org.apache.spark.streaming.kafka010.KafkaUtils;
import org.apache.spark.streaming.kafka010.LocationStrategies;
import scala.Tuple2;

public class SparkStreamingKafkaReceiverExample {
    public static void main(String[] args) throws InterruptedException {
        // 创建 Spark Streaming 上下文
        SparkConf conf = new SparkConf().setAppName("SparkStreamingKafkaReceiverExample");
        JavaStreamingContext jssc = new JavaStreamingContext(conf, Durations.seconds(1));

        // 配置 Kafka 参数
        Map<String, Object> kafkaParams = new HashMap<>();
        kafkaParams.put("bootstrap.servers", "localhost:9092");
        kafkaParams.put("key.deserializer", StringDeserializer.class);
        kafkaParams.put("value.deserializer", StringDeserializer.class);
        kafkaParams.put("group.id", "test-group");
        kafkaParams.put("auto.offset.reset", "latest");
        kafkaParams.put("enable.auto.commit", false);

        // 创建 DStream,从 Kafka 读取数据
        JavaInputDStream<ConsumerRecord<String, String>> kafkaStream = KafkaUtils.createDirectStream(
                jssc,
                LocationStrategies.PreferConsistent(),
                ConsumerStrategies.<String, String>Subscribe(Arrays.asList("test-topic"), kafkaParams)
        );

        // 将每行数据拆分为单词
        JavaDStream<String> lines = kafkaStream.map(record -> record.value());
        JavaDStream<String> words = lines.flatMap(line -> Arrays.asList(line.split(" ")).iterator());

        // 计算每个单词出现的次数
        JavaPairDStream<String, Integer> wordCounts = words.mapToPair(word -> new Tuple2<>(word, 1))
                .reduceByKey((count1, count2) -> count1 + count2);

        // 输出计算结果
        wordCounts.print();

        // 启动 Spark Streaming 应用程序
        jssc.start();
        jssc.awaitTermination();
    }
}
```

示例中首先创建了一个 Spark Streaming 上下文,并配置了 Kafka 参数。然后,使用 createDirectStream() 方法创建了一个 DStream,从 Kafka 读取数据。接着,将每行数据拆分为单词,并使用 mapToPair() 操作将每个单词映射为一个键值对,其中键为单词本身,值为 1,然后使用 reduceByKey() 操作计算每个单词出现的次数。最后,使用 print() 方法输出计算结果。

2. direct 方式

使用 Java 语言编写 Spark Streaming,使用 direct 方式接收 Kafka 数据,示例代码如下。

```java
import java.util.Arrays;
import java.util.HashMap;
import java.util.Map;

import org.apache.kafka.common.serialization.StringDeserializer;
import org.apache.spark.SparkConf;
import org.apache.spark.streaming.Durations;
import org.apache.spark.streaming.api.java.JavaInputDStream;
import org.apache.spark.streaming.api.java.JavaPairDStream;
import org.apache.spark.streaming.api.java.JavaStreamingContext;
import org.apache.spark.streaming.kafka010.ConsumerStrategies;
import org.apache.spark.streaming.kafka010.KafkaUtils;
import org.apache.spark.streaming.kafka010.LocationStrategies;
import scala.Tuple2;

public class SparkStreamingKafkaDirectExample {
    public static void main(String[] args) throws InterruptedException {
        // 创建 Spark Streaming 上下文
        SparkConf conf = new SparkConf().setAppName("SparkStreamingKafkaDirectExample");
        JavaStreamingContext jssc = new JavaStreamingContext(conf, Durations.seconds(1));

        // 配置 Kafka 参数
        Map<String, Object> kafkaParams = new HashMap<>();
        kafkaParams.put("bootstrap.servers", "localhost:9092");
        kafkaParams.put("key.deserializer", StringDeserializer.class);
        kafkaParams.put("value.deserializer", StringDeserializer.class);
        kafkaParams.put("group.id", "test-group");
        kafkaParams.put("auto.offset.reset", "latest");
        kafkaParams.put("enable.auto.commit", false);

        // 创建 DStream,从 Kafka 读取数据
        JavaInputDStream<ConsumerRecord<String, String>> kafkaStream = KafkaUtils.createDirectStream(
            jssc,
            LocationStrategies.PreferConsistent(),
            ConsumerStrategies.<String, String>Subscribe(Arrays.asList("test-topic"), kafkaParams)
        );
```

```java
// 将每行数据拆分为单词
JavaDStream<String> lines = kafkaStream.map(record -> record.value());
JavaDStream<String> words = lines.flatMap(line -> Arrays.asList(line.split(" ")).iterator());

// 计算每个单词出现的次数
JavaPairDStream<String, Integer> wordCounts = words.mapToPair(word -> new Tuple2<>(word, 1))
        .reduceByKey((count1, count2) -> count1 + count2);

// 输出计算结果
wordCounts.print();

// 启动 Spark Streaming 应用程序
jssc.start();
jssc.awaitTermination();
    }
}
```

示例中首先创建了一个 Spark Streaming 上下文,并配置了 Kafka 参数。然后,使用 createDirectStream( ) 方法创建了一个 DStream,从 Kafka 读取数据。接着,将每行数据拆分为单词,并使用 mapToPair( ) 操作将每个单词映射为一个键值对,其中键为单词本身,值为 1,然后使用 reduceByKey( ) 操作计算每个单词出现的次数。最后,使用 print( ) 方法输出计算结果。

第10章 全文搜索引擎 Elasticsearch

本章内容介绍 Elasticsearch 的基本概念，包括索引（index）、类型（type）、文档（document）、字段（field）等，这些概念是构建和操作数据的基础；学习 Elasticsearch 的工作原理，包括倒排索引、分布式架构、数据节点等；学习 Elasticsearch 的安装和配置，集群模式的部署；掌握 Elasticsearch 的常用命令和操作，包括索引创建、文档插入、查询等操作；学习 Elasticsearch 的高级特性和应用场景，包括数据聚合、分析、搜索建议等功能；实践 Elasticsearch 的应用案例，包括日志分析、全文搜索等场景。

10.1 Elasticsearch 简介

1. Elasticsearch 概述

Elasticsearch 是开源分布式搜索引擎，提供搜集、分析、存储数据三大功能。它在内部使用 Luence 做索引与搜索，通过对 Luence 的封装，提供了一套简单一致的 RESTful API。

Elasticsearc 的特点有：分布式、零配置、自动发现、索引自动分片、索引副本机制、RESTful 风格接口、多数据源、自动搜索负载等。

2. Elasticsearch 的特点

Elasticsearch 具有高性能、可扩展性和实时性等特点，同时，支持全文搜索、地理位置搜索和多语言搜索等功能，可以满足各种搜索需求。以下是 Elasticsearch 的特点。

1）分布式架构：Elasticsearch 是一个分布式搜索引擎，可以在多个节点上运行，实现数据的分布式存储和搜索。这使得 Elasticsearch 可以处理大量数据，并提供高可用性和可伸缩性。

2）实时搜索：Elasticsearch 可以实时搜索数据，支持快速的搜索和分析。它可以在毫秒级别内返回搜索结果，并支持实时更新和删除数据。

3）多种查询方式：Elasticsearch 支持多种查询方式，包括全文搜索、结构化搜索、模糊搜索、范围搜索等。它还支持聚合查询、地理位置查询、多语言搜索等高级查询功能。

4）易于扩展：Elasticsearch 可以通过添加新的节点来扩展搜索能力，也可以通过添加新的插件来扩展功能。它还提供了 RESTful API 和 Java API，方便开发人员进行集成和扩展。

5）数据安全：Elasticsearch 提供了多种安全机制，包括身份验证、访问控制、加密传输等，保障数据的安全性和隐私性。

3. Elasticsearch 应用场景

Elasticsearch 广泛应用于日志分析、社交媒体分析、电子商务、搜索引擎、安全监控等领域。它可以用于对日志数据进行实时分析，帮助企业进行故障排除和性能优化。同时，它还可以用于社交媒体分析，对用户行为和趋势进行挖掘。在电子商务领域，Elasticsearch 可以用于产品搜索和推荐，提高用户体验。以下是 Elasticsearch 的应用场景。

1）企业搜索：Elasticsearch 可以用于构建企业搜索引擎，帮助用户快速地搜索和查找企业内部的各种信息，例如文档、邮件、聊天记录等。Elasticsearch 支持全文搜索、模糊搜索、聚合搜索等多种搜索方式，可以满足不同用户的需求。

2）日志分析：Elasticsearch 可以用于处理大量的日志数据，例如服务器日志、应用程序日志等。通过将日志数据存储在 Elasticsearch 中，并使用 Kibana 等工具进行可视化分析，可以帮助用户快速地发现潜在的问题和异常情况。

3）数据分析：Elasticsearch 可以用于处理各种类型的数据，例如销售数据、用户数据等。通过使用 Elasticsearch 的聚合功能，可以对数据进行分组、统计、计算等操作，帮助用户发现数据中的规律和趋势。

4）地理信息系统：Elasticsearch 可以用于处理地理信息数据，例如地图数据、位置数据等。通过使用 Elasticsearch 的地理位置搜索功能，可以帮助用户快速地搜索和查找与地理位置相关的信息。

总之，Elasticsearch 是一款功能强大的搜索引擎，可以应用于各种场景，帮助用户快速地搜索、分析和处理各种类型的数据。

10.2　Elasticsearch 架构和原理

Elasticsearch 的架构组成，包括节点（Nodes）、集群（Cluster）、索引（Index）、分片（Shard）、复制（Replication）等。理解 Elasticsearch 的架构组成，可以帮助我们更好地设计、部署和优化 Elasticsearch 集群。Elasticsearch 的架构组成如图 10-1 所示。

节点是 Elasticsearch 的基本单元，可以是主节点或数据节点。主节点负责管理集群的全局状态，包括索引和分片的分配，以及节点的加入和退出。数据节点则负责存储和处理数据，包括索引和查询操作。

1. 主节点的作用

1）管理集群的全局状态，包括索引和分片的分配，以及节点的加入和退出。
2）处理集群级别的操作，如创建和删除索引，设置索引级别的参数等。
3）协调分片的分配和复制，以确保数据的可用性和一致性。
4）选举新的主节点，以防止主节点故障或失效。

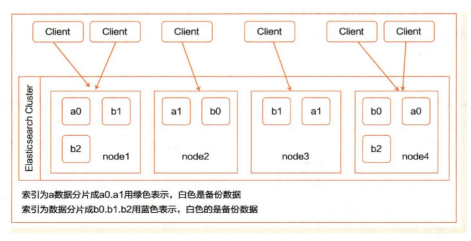

· 图 10-1 Elasticsearch 架构

2. 数据节点的作用

1）存储和处理数据，包括索引和查询操作。
2）执行搜索请求，返回匹配查询条件的文档。
3）处理分片级别的操作，如创建和删除文档，更新和删除文档等。
4）处理本地的聚合查询，返回分片级别的聚合结果。

在 Elasticsearch 集群中，每个节点可以同时扮演主节点和数据节点的角色，也可以只扮演其中一种角色。主节点和数据节点的数量和分布方式可以根据实际需求进行配置和调整，以提高集群的可用性和性能。

10.2.1 Elasticsearch 核心概念

Elasticsearch 核心概念主要分两大部分，第一部分是与节点相关的核心概念，第二部分是与文档相关的核心概念。

1. 集群、节点、分片分布式三要素

1）Cluster（集群）：集群中有多个节点（Node），其中有一个为主节点，这个主节点是可以通过选举产生的，主从节点是对于集群内部来说的。Elasticsearch 的一个概念就是去中心化，字面上理解就是无中心节点，这是对于集群外部来说的，你与任何一个节点的通信和与整个 Elasticsearch 集群通信是等价的。

2）Node（节点）：一个运行中的 Elasticsearch 实例称为一个节点，一个节点是集群中的一个服务器，作为集群的一部分，它存储数据，参与集群的索引和搜索功能。一个节点也是由一个名字来标识的。一个节点可以通过配置集群名称的方式来加入一个指定的集群。

3）Shards（分片）：代表索引分片，Elasticsearch 可以把一个完整的索引分成多个分片，这样的好处是可以把一个大的索引拆分成多个，分布到不同的节点上。构成分布式搜索。分片的数量只能在索引创建前指定，并且索引创建后不能更改。

4）Replicas（副本）：代表索引副本，Elasticsearch 可以设置多个索引的副本，副本的作用一是提高系统的容错性，当某个节点某个分片损坏或丢失时，可以从副本中恢复；二是提高 Elasticsearch 的查询效率，Elasticsearch 会自动对搜索请求进行负载均衡。

2. 索引、类型、文档三要素

1）Index（索引-数据库）：Elasticsearch 将它的数据存储在一个或多个索引（Index）中。用 SQL 领域的术语来类比，索引就像数据库，可以向索引写入文档或者从索引中读取文档，并通过 Elasticsearch 内部使用的 Luence 将数据写入索引或从索引中检索数据。

2）Type（类型-表）：每个文档都有与之对应的类型（Type）定义。这允许用户在一个索引中存储多种文档类型，并为不同文档类型提供不同的映射。

3）Document（文档）：文档是 Elasticsearch 中的最小数据单元，一个 Document 就是一条数据，通常用 JSON 数据结构表示，每个 Index 下的 Type 中，都可以存储多个 Document。

4）Field（字段-列）：Field 是 Elasticsearch 的最小单位。一个 Document 里面有多个 Field，每个 Field 就是一个数据字段。

5）Mapping（映射-约束）：所有文档写进索引之前都会先进行分析，如何将输入的文本分割为词条、哪些词条又会被过滤，这种行为叫作映射（Mapping）。Mapping 用来定义一个文档，可以定义所包含的字段以及字段的类型、分词器及属性等。

10.2.2 Elasticsearch 工作原理

1. Elasticsearch 写入数据的过程

通过理解 Elasticsearch 写入数据的过程，可以更好地设计和优化数据写入操作，以提高系统的性能和吞吐量。Elasticsearch 写入数据的过程，如图 10-2 所示。

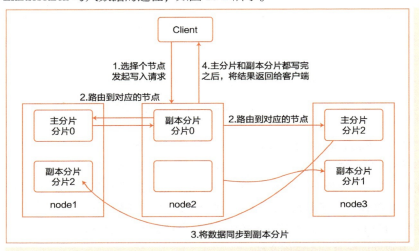

● 图 10-2　Elasticsearch 写入数据的过程

Elasticsearch 写入数据的过程分为以下几个步骤。

1）索引文档：在 Elasticsearch 中，文档被索引到一个或多个索引中。每个索引包含一个或多个分片，每个分片包含一部分文档的数据。当文档被索引时，它会被分配到一个或多个分片中。

2）分析文档：当用户写入一个文档时，Elasticsearch 会对文档进行分析，包括分词、过滤、归一化等操作。这些操作可以帮助 Elasticsearch 更好地理解文档的内容。

3）写入分片：Elasticsearch 会将文档写入一个或多个分片中。每个分片会独立地执行写入操作，并将文档存储在本地磁盘上。

4）等待刷新：当文档被写入分片中后，Elasticsearch 会等待一段时间，直到所有分片完成写入操作，并将文档刷新到磁盘上。这个过程可以通过设置刷新间隔来控制。

在实际应用中，Elasticsearch 还提供了许多高级写入功能，如批量写入、更新、删除等，可以帮助用户更好地管理数据。同时，Elasticsearch 还支持分布式部署，可以通过添加更多的节点来提高写入性能和可用性。

2. Elasticsearch 搜索数据的过程

图 10-3 展示了 Elasticsearch 搜索数据的过程，可以简单概括为以下几个步骤。

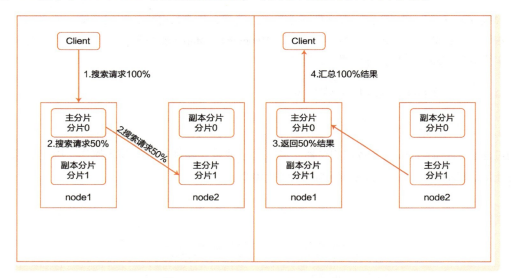

● 图 10-3　Elasticsearch 搜索数据过程

1）索引文档：在 Elasticsearch 中，文档被索引到一个或多个索引中。每个索引包含一个或多个分片，每个分片包含一部分文档的数据。当文档被索引时，它会被分配到一个或多个分片中。

2）分析查询：当用户发起一个查询请求时，Elasticsearch 会对查询进行分析，包括分词、过滤、归一化等操作。这些操作可以帮助 Elasticsearch 更好地理解用户的查询意图。

3）查询分片：Elasticsearch 会将查询请求发送到每个包含相关数据的分片中。每个分片会独立地执行查询操作，并返回匹配查询条件的文档。

4）合并结果：当所有分片返回了查询结果后，Elasticsearch 会将这些结果合并成一个完整的结果集，并按照相关性进行排序。最终，Elasticsearch 会将结果返回给用户。

在实际应用中，Elasticsearch 还提供了许多高级搜索功能，如聚合、过滤器、排序等，可以帮助用户更好地定制搜索结果。同时，Elasticsearch 还支持分布式部署，可以通过添加更多的节点来提高搜索性能和可用性。

10.2.3 Elasticsearch 倒排索引

Elasticsearch 核心是使用一种称为倒排索引的结构，它适用于快速地全文搜索。一个倒排索引由文档中所有不重复词的列表构成，对于其中每个词，有一个包含它的文档列表。

1. 概述

倒排索引是区别于正排索引来说的。

（1）正排索引

正排索引是从文档角度来找其中的单词，表示每个文档都含有哪些单词，以及每个单词出现了多少次（词频）及其出现位置（相对于文档首部的偏移量）。所以每次搜索都是遍历所有文章。

（2）倒排索引

倒排索引是从单词角度找文档，标识每个单词分别在哪些文档中出现（文档 ID），以及在各自的文档中每个单词分别出现了多少次（词频）及其出现位置（相对于该文档首部的偏移量）。

Elasticsearch 倒排索引示意图如图 10-4 所示。

● 图 10-4　Elasticsearch 倒排索引示意图

Elasticsearch 的倒排索引组成如下。

1）单词词典：所有单词的倒排列表顺序存储在磁盘的某个文件里，这个文件即被称为倒排文件，倒排文件是存储倒排索引的物理文件。

2）倒排文件：单词词典是由文档集合中出现过的所有单词构成的字符串集合，单词词典内每条索引项记载单词本身的一些信息以及指向"倒排列表"的指针。

单词词典是倒排索引中非常重要的组成部分，它是用来维护文档集合中所有单词的相关信息，同时用来记载某个单词对应的倒排列表在倒排文件中的位置信息。在支持搜索时，根据用户的查询词，去单词词典里查询，就能够获得相应的倒排列表。

2. 创建倒排索引过程

为了创建倒排索引，我们首先将每个文档拆分成单独的词（称它为词条），创建一个包含所有不重复词条的排序列表，然后列出每个词条出现在哪个文档。

1）分词：Lucene 将上面三列分别作为词典文件、频率文件、位置文件保存。其中词典文件不仅保存有每个关键词，还保留了指向频率文件和位置文件的指针，通过指针可以找到该关键字的频率信息和位置信息。

2）创建文档列表：Lucene 首先对原始文档数据进行编号（DocID），形成列表，就是一个文档列表。

3）创建倒排索引列表：然后对文档中的数据进行分词，得到词条。对词条进行编号，以词条创建索引。然后记录下包含该词条的所有文档编号（及其他信息）。记录文章中出现的次数和出现的位置，通常有如下两种方法。

字符位置：即记录该词是文章中第几个字符（优点是关键词亮显时定位快）。

关键词位置：先把文章进行分词，然后记录该词是文章中第几个关键词（优点是节约索引空间、词组查询快），Lucene 中记录的就是这种方法。

10.3 Elasticsearch 实战

10.3.1 Elasticsearch 索引创建

以下是使用 Elasticsearch 创建索引的代码。

```java
import org.elasticsearch.action.index.IndexRequest;
import org.elasticsearch.action.index.IndexResponse;
import org.elasticsearch.client.RestHighLevelClient;
import org.elasticsearch.common.xcontent.XContentType;
import java.io.IOException;

public class CreateProductIndexExample {
    public static void main(String[] args) throws IOException {
        RestHighLevelClient client = new RestHighLevelClient(
            RestClient.builder(new HttpHost("localhost", 9200, "http")));

        IndexRequest request = new IndexRequest("product_index");
        request.id("product_id");
        String jsonString = "{" +
            "\"name\":\"product_name\"," +
```

```
            "\"description\":\"product_description\"," +
            "\"price\":100.0," +
            "\"category\":\"product_category\"" +
            "}";
        request.source(jsonString, XContentType.JSON);

        IndexResponse response = client.index(request);

        client.close();
    }
}
```

在这个示例中，我们首先创建了一个 RestHighLevelClient 对象，然后创建了一个 IndexRequest 对象，指定了要创建商品索引的索引名称和商品 ID。接下来，创建了一个 JSON 字符串，表示要创建的商品内容，并将其设置为 IndexRequest 的 source。最后，使用 client.index ( ) 方法将商品索引保存到 Elasticsearch 中。

10.3.2　Elasticsearch 索引更新

以下是使用 Elasticsearch 更新索引的代码。

```
import org.elasticsearch.action.update.UpdateRequest;
import org.elasticsearch.action.update.UpdateResponse;
import org.elasticsearch.client.RestHighLevelClient;
import org.elasticsearch.common.xcontent.XContentType;
import java.io.IOException;

public class UpdateProductIndexExample {
    public static void main(String[] args) throws IOException {
        RestHighLevelClient client = new RestHighLevelClient(
            RestClient.builder(new HttpHost("localhost", 9200, "http")));

        UpdateRequest request = new UpdateRequest("product_index", "product_id");
        String jsonString = "{" +
            "\"name\":\"updated_product_name\"," +
            "\"description\":\"updated_product_description\"," +
            "\"price\":200.0," +
            "\"category\":\"updated_product_category\"" +
            "}";
        request.doc(jsonString, XContentType.JSON);

        UpdateResponse response = client.update(request);

        client.close();
    }
}
```

在这个示例中，我们首先创建了一个 RestHighLevelClient 对象，然后创建了一个 UpdateRequest 对象，指定了要更新商品索引的索引名称和商品 ID。接下来，创建了一个 JSON 字符串，表示要更新的商品内容，并将其设置为 UpdateRequest 的 doc。最后，使用 client.update() 方法将商品索引更新到 Elasticsearch 中。

10.3.3 Elasticsearch 索引查询

以下是使用 Elasticsearch 索引查询的代码。

```java
import org.elasticsearch.action.search.SearchRequest;
import org.elasticsearch.action.search.SearchResponse;
import org.elasticsearch.client.RestHighLevelClient;
import org.elasticsearch.common.unit.TimeValue;
import org.elasticsearch.index.query.QueryBuilders;
import org.elasticsearch.search.builder.SearchSourceBuilder;
import java.io.IOException;

public class QueryProductIndexExample {
    public static void main(String[] args) throws IOException {
        RestHighLevelClient client = new RestHighLevelClient(
            RestClient.builder(new HttpHost("localhost", 9200, "http")));

        SearchRequest searchRequest = new SearchRequest("product_index");
        SearchSourceBuilder searchSourceBuilder = new SearchSourceBuilder();
        searchSourceBuilder.query(QueryBuilders.matchQuery("name", "product_name"));
        searchSourceBuilder.from(0);
        searchSourceBuilder.size(5);
        searchSourceBuilder.timeout(new TimeValue(60, TimeUnit.SECONDS));
        searchRequest.source(searchSourceBuilder);

        SearchResponse searchResponse = client.search(searchRequest);

        client.close();
    }
}
```

在这个示例中，我们首先创建了一个 RestHighLevelClient 对象，然后创建了一个 SearchRequest 对象，指定了要查询商品索引的索引名称。接下来，创建了一个 SearchSourceBuilder 对象，并将查询条件设置为匹配商品名称。我们还设置了 from 和 size 参数来限制返回结果的数量，并设置了一个 60 秒的超时时间。最后，将 SearchSourceBuilder 对象设置为 SearchRequest 的 source，并使用 client.search() 方法执行查询。

10.3.4 Elasticsearch 索引删除

以下是使用 Elasticsearch 索引删除的代码。

```
import org.elasticsearch.action.delete.DeleteRequest;
import org.elasticsearch.action.delete.DeleteResponse;
import org.elasticsearch.client.RestHighLevelClient;
import java.io.IOException;

public class DeleteProductIndexExample {
    public static void main(String[] args) throws IOException {
        RestHighLevelClient client = new RestHighLevelClient(
            RestClient.builder(new HttpHost("localhost", 9200, "http")));

        DeleteRequest request = new DeleteRequest("product_index", "product_id");

        DeleteResponse response = client.delete(request);

        client.close();
    }
}
```

在这个示例中，我们首先创建了一个 RestHighLevelClient 对象，然后创建了一个 DeleteRequest 对象，指定了要删除商品索引的索引名称和商品 ID。最后，使用 client.delete() 方法将商品索引从 Elasticsearch 中删除。

10.3.5　Elasticsearch 保存文档

以下是使用 Elasticsearch 保存文档的代码。

```
import org.elasticsearch.action.index.IndexRequest;
import org.elasticsearch.action.index.IndexResponse;
import org.elasticsearch.client.RestHighLevelClient;
import org.elasticsearch.common.xcontent.XContentType;
import java.io.IOException;

public class IndexProductExample {
    public static void main(String[] args) throws IOException {
        RestHighLevelClient client = new RestHighLevelClient(
            RestClient.builder(new HttpHost("localhost", 9200, "http")));

        String index = "product_index";
        String type = "product_type";
        String id = "product_id";
        String jsonString = "{" +
            "\"name\":\"product_name\"," +
            "\"price\":100.0," +
            "\"description\":\"product_description\"" +
            "}";
```

```
        IndexRequest request = new IndexRequest(index, type, id);
        request.source(jsonString, XContentType.JSON);

        IndexResponse response = client.index(request);

        client.close();
    }
}
```

在这个示例中，我们首先创建了一个 RestHighLevelClient 对象，然后定义了商品索引的名称、类型和 ID。接下来，创建了一个 JSON 字符串，包含了商品的名称、价格和描述信息。然后，创建了一个 IndexRequest 对象，并将 JSON 字符串作为 source 添加到请求中。最后，使用 client.index( ) 方法将商品索引添加到 Elasticsearch 中。

10.3.6 Elasticsearch 更新文档

以下是使用 Elasticsearch 更新文档的代码。

```
import org.elasticsearch.action.update.UpdateRequest;
import org.elasticsearch.action.update.UpdateResponse;
import org.elasticsearch.client.RestHighLevelClient;
import org.elasticsearch.common.xcontent.XContentType;
import java.io.IOException;

public class UpdateProductExample {
    public static void main(String[] args) throws IOException {
        RestHighLevelClient client = new RestHighLevelClient(
            RestClient.builder(new HttpHost("localhost", 9200, "http")));

        String index = "product_index";
        String type = "product_type";
        String id = "product_id";
        String jsonString = "{" +
            "\"name\":\"updated_product_name\"," +
            "\"price\":200.0," +
            "\"description\":\"updated_product_description\"" +
            "}";

        UpdateRequest request = new UpdateRequest(index, type, id);
        request.doc(jsonString, XContentType.JSON);

        UpdateResponse response = client.update(request);

        client.close();
    }
}
```

在这个示例中，我们首先创建了一个 RestHighLevelClient 对象，然后定义了商品文档所在的索引名称、类型和 ID。接下来，创建了一个 JSON 字符串，包含了更新后的商品的名称、价格和描述信息。然后，创建了一个 UpdateRequest 对象，并将 JSON 字符串作为 doc 添加到请求中。最后，使用 client.update() 方法将更新后的商品文档添加到 Elasticsearch 中。

10.3.7　Elasticsearch 精确查询

以下是使用 Elasticsearch 精确查询文档的代码。

```java
import org.elasticsearch.action.search.SearchRequest;
import org.elasticsearch.action.search.SearchResponse;
import org.elasticsearch.client.RestHighLevelClient;
import org.elasticsearch.common.unit.Fuzziness;
import org.elasticsearch.common.unit.TimeValue;
import org.elasticsearch.index.query.QueryBuilders;
import org.elasticsearch.search.builder.SearchSourceBuilder;
import java.io.IOException;

public class ExactSearchProductExample {
    public static void main(String[] args) throws IOException {
        RestHighLevelClient client = new RestHighLevelClient(
            RestClient.builder(new HttpHost("localhost", 9200, "http")));

        String index = "product_index";
        String type = "product_type";
        String field = "name";
        String value = "product1";

        SearchRequest searchRequest = new SearchRequest(index);
        searchRequest.types(type);

        SearchSourceBuilder searchSourceBuilder = new SearchSourceBuilder();
        searchSourceBuilder.query(QueryBuilders.matchQuery(field, value).fuzziness(Fuzziness.ZERO));
        searchSourceBuilder.timeout(new TimeValue(60, TimeUnit.SECONDS));
        searchRequest.source(searchSourceBuilder);

        SearchResponse searchResponse = client.search(searchRequest);

        client.close();
    }
}
```

在这个示例中，我们首先创建了一个 RestHighLevelClient 对象，然后定义了商品文档所在的索引名称和类型，以及要查询的字段名称和字段值。接下来，创建了一个 SearchRequest 对象，并将索引名称和类型作为参数传入。然后，创建了一个 SearchSourceBuilder 对象，并使用 QueryBuilders.matchQuery()

方法创建了一个匹配查询，将字段名称和字段值作为参数传入，并使用 fuzziness（Fuzziness.ZERO）方法设置了精确匹配。最后，将 SearchSourceBuilder 对象作为参数传入 SearchRequest 对象，并使用 client.search() 方法执行查询操作。

10.3.8 Elasticsearch 模糊查询

以下是使用 Elasticsearch 模糊查询文档的代码。

```java
import org.elasticsearch.action.search.SearchRequest;
import org.elasticsearch.action.search.SearchResponse;
import org.elasticsearch.client.RestHighLevelClient;
import org.elasticsearch.common.unit.Fuzziness;
import org.elasticsearch.common.unit.TimeValue;
import org.elasticsearch.index.query.QueryBuilders;
import org.elasticsearch.search.builder.SearchSourceBuilder;
import java.io.IOException;

public class FuzzySearchProductExample {
    public static void main(String[] args) throws IOException {
        RestHighLevelClient client = new RestHighLevelClient(
            RestClient.builder(new HttpHost("localhost", 9200, "http")));

        String index = "product_index";
        String type = "product_type";
        String field = "name";
        String value = "product1";

        SearchRequest searchRequest = new SearchRequest(index);
        searchRequest.types(type);

        SearchSourceBuilder searchSourceBuilder = new SearchSourceBuilder();
        searchSourceBuilder.query(QueryBuilders.matchQuery(field, value).fuzziness(Fuzziness.AUTO));
        searchSourceBuilder.timeout(new TimeValue(60, TimeUnit.SECONDS));
        searchRequest.source(searchSourceBuilder);

        SearchResponse searchResponse = client.search(searchRequest);

        client.close();
    }
}
```

在这个示例中，我们首先创建了一个 RestHighLevelClient 对象，然后定义了商品文档所在的索引名称和类型，以及要查询的字段名称和字段值。接下来，创建了一个 SearchRequest 对象，并将索引名称和类型作为参数传入。然后，创建了一个 SearchSourceBuilder 对象，并使用 QueryBuilders.matchQuery() 方法创建了一个匹配查询，将字段名称和字段值作为参数传入，并使用 fuzziness（Fuzziness.AUTO）方法设置了模糊匹配。最后，将 SearchSourceBuilder 对象作为参数传入 SearchRequest 对象，并使用 client.search( ) 方法执行查询操作。

▶▶ 10.3.9　Elasticsearch 范围查询

以下是使用 Elasticsearch 范围查询文档的代码。

```java
import org.elasticsearch.action.search.SearchRequest;
import org.elasticsearch.action.search.SearchResponse;
import org.elasticsearch.client.RestHighLevelClient;
import org.elasticsearch.common.unit.TimeValue;
import org.elasticsearch.index.query.QueryBuilders;
import org.elasticsearch.search.builder.SearchSourceBuilder;
import java.io.IOException;

public class RangeSearchProductExample {
    public static void main(String[] args) throws IOException {
        RestHighLevelClient client = new RestHighLevelClient(
            RestClient.builder(new HttpHost("localhost", 9200, "http")));

        String index = "product_index";
        String type = "product_type";
        String field = "price";
        int minPrice = 10;
        int maxPrice = 100;

        SearchRequest searchRequest = new SearchRequest(index);
        searchRequest.types(type);

        SearchSourceBuilder searchSourceBuilder = new SearchSourceBuilder();
        searchSourceBuilder.query(QueryBuilders.rangeQuery(field).gte(minPrice).lte(maxPrice));
        searchSourceBuilder.timeout(new TimeValue(60, TimeUnit.SECONDS));
        searchRequest.source(searchSourceBuilder);

        SearchResponse searchResponse = client.search(searchRequest);

        client.close();
    }
}
```

在这个示例中，我们首先创建了一个 RestHighLevelClient 对象，然后定义了商品文档所在的索引名称和类型，以及要查询的字段名称和范围。接下来，创建了一个 SearchRequest 对象，并将索引名称和类型作为参数传入。然后，创建了一个 SearchSourceBuilder 对象，并使用 QueryBuilders.rangeQuery( ) 方法创建了一个范围查询，将字段名称和范围作为参数传入。最后，将 SearchSourceBuilder 对象作为参数传入 SearchRequest 对象，并使用 client.search( ) 方法执行查询操作。

10.3.10　Elasticsearch 布尔查询

以下是使用 Elasticsearch 布尔查询文档的代码。

```java
import org.elasticsearch.action.search.SearchRequest;
import org.elasticsearch.action.search.SearchResponse;
import org.elasticsearch.client.RestHighLevelClient;
import org.elasticsearch.common.unit.TimeValue;
import org.elasticsearch.index.query.BoolQueryBuilder;
import org.elasticsearch.index.query.QueryBuilders;
import org.elasticsearch.search.builder.SearchSourceBuilder;
import java.io.IOException;

public class BoolSearchProductExample {
    public static void main(String[] args) throws IOException {
        RestHighLevelClient client = new RestHighLevelClient(
            RestClient.builder(new HttpHost("localhost", 9200, "http")));

        String index = "product_index";
        String type = "product_type";
        String field1 = "name";
        String value1 = "prod* ";
        String field2 = "price";
        int minPrice = 10;
        int maxPrice = 100;

        SearchRequest searchRequest = new SearchRequest(index);
        searchRequest.types(type);

        SearchSourceBuilder searchSourceBuilder = new SearchSourceBuilder();
        BoolQueryBuilder boolQueryBuilder = QueryBuilders.boolQuery();
        boolQueryBuilder.must(QueryBuilders.wildcardQuery(field1, value1));
        boolQueryBuilder.filter(QueryBuilders.rangeQuery(field2).gte(minPrice).lte(maxPrice));
        searchSourceBuilder.query(boolQueryBuilder);
        searchSourceBuilder.timeout(new TimeValue(60, TimeUnit.SECONDS));
        searchRequest.source(searchSourceBuilder);

        SearchResponse searchResponse = client.search(searchRequest);

        client.close();
    }
}
```

在这个示例中，我们首先创建了一个 RestHighLevelClient 对象，然后定义了商品文档所在的索引名称和类型，以及要查询的字段名称和布尔查询条件。接下来，创建了一个 SearchRequest 对象，并将索引名称和类型作为参数传入。然后，创建了一个 SearchSourceBuilder 对象，并使用 QueryBuilders.boolQuery()

方法创建了一个布尔查询，将多个查询条件作为参数传入。其中，我们使用 must( ) 方法添加了一个通配查询条件，使用 filter( ) 方法添加了一个范围查询条件。最后，将 SearchSourceBuilder 对象作为参数传入 SearchRequest 对象，并使用 client.search( ) 方法执行查询操作。

10.3.11　Elasticsearch 聚合查询

以下是使用 Elasticsearch 聚合查询文档的代码。

```java
import org.elasticsearch.action.search.SearchRequest;
import org.elasticsearch.action.search.SearchResponse;
import org.elasticsearch.client.RestHighLevelClient;
import org.elasticsearch.common.unit.TimeValue;
import org.elasticsearch.index.query.QueryBuilders;
import org.elasticsearch.search.aggregations.AggregationBuilders;
import org.elasticsearch.search.aggregations.metrics.AvgAggregationBuilder;
import org.elasticsearch.search.builder.SearchSourceBuilder;
import java.io.IOException;

public class AvgAggregationSearchProductExample {
    public static void main(String[] args) throws IOException {
        RestHighLevelClient client = new RestHighLevelClient(
            RestClient.builder(new HttpHost("localhost", 9200, "http")));

        String index = "product_index";
        String type = "product_type";

        SearchRequest searchRequest = new SearchRequest(index);
        searchRequest.types(type);

        SearchSourceBuilder searchSourceBuilder = new SearchSourceBuilder();

        AvgAggregationBuilder aggregationBuilder = AggregationBuilders.avg("avg_price").field("price");

        searchSourceBuilder.aggregation(aggregationBuilder);
        searchSourceBuilder.timeout(new TimeValue(60, TimeUnit.SECONDS));
        searchRequest.source(searchSourceBuilder);

        SearchResponse searchResponse = client.search(searchRequest);

        client.close();
    }
}
```

在这个示例中，我们首先创建了一个 RestHighLevelClient 对象，然后定义了商品文档所在的索引名称和类型。接下来，创建了一个 SearchRequest 对象，并将索引名称和类型作为参数传入。然后,创建了一个 SearchSourceBuilder 对象,并使用 AggregationBuilders.avg() 方法创建了一个平均值聚合查询。在平均值聚合查询中,我们使用"price"作为聚合字段,表示按照商品价格进行聚合。最后,将 AvgAggregationBuilder 对象作为参数传入 SearchSourceBuilder 对象,并使用 client.search() 方法执行查询。

▶▶ 10.3.12 Elasticsearch 高亮查询

以下是使用 Elasticsearch 高亮查询文档的代码。

```java
import org.elasticsearch.action.search.SearchRequest;
import org.elasticsearch.action.search.SearchResponse;
import org.elasticsearch.client.RestHighLevelClient;
import org.elasticsearch.common.unit.TimeValue;
import org.elasticsearch.index.query.QueryBuilders;
import org.elasticsearch.search.SearchHit;
import org.elasticsearch.search.builder.SearchSourceBuilder;
import org.elasticsearch.search.fetch.subphase.highlight.HighlightBuilder;
import org.elasticsearch.search.fetch.subphase.highlight.HighlightField;
import java.io.IOException;
import java.util.Map;

public class HighlightSearchProductExample {
    public static void main(String[] args) throws IOException {
        RestHighLevelClient client = new RestHighLevelClient(
            RestClient.builder(new HttpHost("localhost", 9200, "http")));

        String index = "product_index";
        String type = "product_type";

        SearchRequest searchRequest = new SearchRequest(index);
        searchRequest.types(type);

        SearchSourceBuilder searchSourceBuilder = new SearchSourceBuilder();

        HighlightBuilder highlightBuilder = new HighlightBuilder();
        highlightBuilder.field("name");
        highlightBuilder.field("description");

        searchSourceBuilder.query(QueryBuilders.matchQuery("name", "iphone"));
        searchSourceBuilder.highlighter(highlightBuilder);
        searchSourceBuilder.timeout(new TimeValue(60, TimeUnit.SECONDS));
        searchRequest.source(searchSourceBuilder);

        SearchResponse searchResponse = client.search(searchRequest);

        SearchHit[] searchHits = searchResponse.getHits().getHits();
```

```
        for (SearchHit hit : searchHits) {
            Map<String, HighlightField> highlightFields = hit.getHighlightFields();
            HighlightField nameField = highlightFields.get("name");
            HighlightField descriptionField = highlightFields.get("description");
            String name = nameField.getFragments()[0].toString();
            String description = descriptionField.getFragments()[0].toString();
            System.out.println("name: " + name);
            System.out.println("description: " + description);
        }

        client.close();
    }
}
```

在这个示例中，我们首先创建了一个 RestHighLevelClient 对象，然后定义了商品文档所在的索引名称和类型。接下来，创建了一个 SearchRequest 对象，并将索引名称和类型作为参数传入。然后，创建了一个 SearchSourceBuilder 对象，并使用 QueryBuilders.matchQuery() 方法创建了一个匹配查询，表示查询商品名称中包含 "iphone" 的商品。接下来，创建了一个 HighlightBuilder 对象，并使用 highlightBuilder.field() 方法指定了需要高亮的字段。最后，将 HighlightBuilder 对象作为参数传入 SearchSourceBuilder 对象，并使用 client.search() 方法执行查询操作。

10.4 Elasticsearch 实现搜索系统

Elasticsearch 是一个分布式、可扩展、实时的搜索与数据分析引擎，通过它我们可以构建出一个强大的全文搜索系统。

搜索系统的作用，就是帮助用户在大量的信息中找到自己喜欢的信息，从而减少用户查找信息的时间。比如我们平时搜索信息的时候用百度搜索，或者是购买商品的时候，在淘宝、京东商场搜索自己喜欢的商品。

搜索系统能够根据用户输入的关键字匹配并且通过评分来进行排序，能够最大程度地帮助用户搜索想要的信息，极大地提高用户的效率。

▶ 10.4.1 搜索系统项目环境准备

在这个教程中，我们将介绍如何为搜索系统项目做好充分的环境准备。通过本实战，你将了解搜索系统的基本原理、所需技术和工具，以及如何搭建一个稳定、高效的项目环境。

1. 创建搜索项目并且导入依赖

创建一个搜索项目，命名为 jareny-bigdata-elasticsearch-repository，并且在项目的 pom.xml 文件中导入相关的依赖，实例代码如下。

```
<?xml version="1.0" encoding="UTF-8"?>
<project xmlns="http://maven.apache.org/POM/4.0.0" xmlns:xsi="http://www.w3.org/2001/
XMLSchema-instance"
```

```xml
        xsi:schemaLocation="http://maven.apache.org/POM/4.0.0 https://maven.apache.org/xsd/maven-4.0.0.xsd">
    <modelVersion>4.0.0</modelVersion>
    <parent>
        <groupId>org.springframework.boot</groupId>
        <artifactId>spring-boot-starter-parent</artifactId>
        <version>2.4.0</version>
    </parent>

    <groupId>com.it.jareny.bigdata</groupId>
    <artifactId>jareny-bigdata-elasticsearch-repository</artifactId>
    <version>0.0.1-SNAPSHOT</version>
    <name>jareny-bigdata-elasticsearch-repository</name>
    <description>jareny-bigdata-elasticsearch-repository</description>

    <properties>
        <java.version>1.8</java.version>
        <!--自定义 es 版本依赖,保证和本地一致-->
        <elasticsearch.version>7.3.2</elasticsearch.version>
    </properties>

    <dependencies>
        <dependency>
            <groupId>org.springframework.boot</groupId>
            <artifactId>spring-boot-starter-web</artifactId>
        </dependency>
        <dependency>
            <groupId>org.springframework.boot</groupId>
            <artifactId>spring-boot-starter-data-elasticsearch</artifactId>
        </dependency>

        <dependency>
            <groupId>org.jsoup</groupId>
            <artifactId>jsoup</artifactId>
            <version>1.10.2</version>
        </dependency>
        <dependency>
            <groupId>com.alibaba</groupId>
            <artifactId>fastjson</artifactId>
            <version>1.2.31</version>
        </dependency>

        <dependency>
            <groupId>org.springframework.boot</groupId>
            <artifactId>spring-boot-devtools</artifactId>
            <scope>runtime</scope>
```

```xml
            <optional>true</optional>
        </dependency>
        <dependency>
            <groupId>org.projectlombok</groupId>
            <artifactId>lombok</artifactId>
            <optional>true</optional>
        </dependency>
        <dependency>
            <groupId>org.springframework.boot</groupId>
            <artifactId>spring-boot-starter-test</artifactId>
            <scope>test</scope>
        </dependency>
    </dependencies>

    <build>
        <plugins>
            <plugin>
                <groupId>org.springframework.boot</groupId>
                <artifactId>spring-boot-maven-plugin</artifactId>
            </plugin>
        </plugins>
    </build>
</project>
```

2. 编写启动类

创建一个启动类，命名为 ElasticsearchRepository，用来启动搜索的项目，实例代码如下。

```
package com.it.jareny.bigdata.elasticsearch;

import org.springframework.boot.SpringApplication;
import org.springframework.boot.autoconfigure.SpringBootApplication;
import org.springframework.data.elasticsearch.repository.config.EnableElasticsearchRepositories;

@EnableElasticsearchRepositories(basePackages = "com.it.jareny.bigdata.elasticsearch.repository")
@SpringBootApplication
public class ElasticsearchRepository {

    public static void main(String[] args) {
        SpringApplication.run(ElasticsearchRepository.class, args);
    }
}
```

3. 添加 Elasticsearch 配置信息

在项目的/src/main/resources 下新建一个 application.properties 文件，用于项目配置。实例代码如下。

```
server.port=9090
server.servlet.context-path=/es
#如果是集群,用逗号隔开
elasticsearch.address=192.168.81.111:9200,192.168.81.112:9200,192.168.81.113:9200
#连接超时时间
elasticsearch.connect-timeout=1000
#连接超时时间
elasticsearch.socket-timeout=30000
#获取连接的超时时间
elasticsearch.connection-request-timeout=500
#最大连接数
elasticsearch.max-connect-num=100
#最大路由连接数
elasticsearch.max-connect-per-route=100
```

4. 创建 Elasticsearch 配置类

新建一个 Elasticsearch 配置类,用于配置 Elasticsearch 的连接,实例代码如下。

```
package com.it.jareny.bigdata.elasticsearch.config;

import org.apache.http.HttpHost;
import org.apache.http.client.config.RequestConfig;
import org.apache.http.impl.nio.client.HttpAsyncClientBuilder;
import org.elasticsearch.client.RestClient;
import org.elasticsearch.client.RestClientBuilder;
import org.elasticsearch.client.RestHighLevelClient;
import org.springframework.beans.factory.annotation.Value;
import org.springframework.context.annotation.Bean;
import org.springframework.context.annotation.Configuration;
import org.springframework.data.elasticsearch.config.AbstractElasticsearchConfiguration;
import java.util.ArrayList;

@Configuration
public class RestClientConfig extends AbstractElasticsearchConfiguration {
    // 创建 ES 连接地址
    @Value("${elasticsearch.address}")
    private String address;

    // 连接超时时间
    @Value("${elasticsearch.connect-timeout}")
    private int connectTimeOut = 1000;

    // 连接超时时间
    @Value("${elasticsearch.socket-timeout}")
```

```java
private int socketTimeOut = 30000;

//   获取连接的超时时间
@Value("${elasticsearch.connection-request-timeout}")
private int connectionRequestTimeOut = 500;

// 最大连接数
@Value("${elasticsearch.max-connect-num}")
private int maxConnectNum = 100;

// 最大路由连接数
@Value("${elasticsearch.max-connect-per-route}")
private int maxConnectPerRoute = 100;

@Bean
@Override
public RestHighLevelClient elasticsearchClient() {
    ArrayList<HttpHost> hostList = new ArrayList<>();
    String[] addrss = address.split(",");
    for(String addr : addrss){
        String[] arr = addr.split(":");
        hostList.add(new HttpHost(arr[0],
                Integer.parseInt(arr[1]),
                "http"));
    }

    RestClientBuilder builder = RestClient.builder(
            new HttpHost("192.168.81.111", 9200, "http"),
            new HttpHost("192.168.81.112", 9200, "http"),
            new HttpHost("192.168.81.113", 9200, "http"));
    // 异步 httpclient 连接延时配置
    builder.setRequestConfigCallback(new RestClientBuilder.RequestConfigCallback() {
        @Override
        public RequestConfig.Builder customizeRequestConfig(
                RequestConfig.Builder requestConfigBuilder) {

            requestConfigBuilder.setConnectTimeout(connectTimeOut);
            requestConfigBuilder.setSocketTimeout(socketTimeOut);
            requestConfigBuilder.setConnectionRequestTimeout(
                    connectionRequestTimeOut);
            return requestConfigBuilder;
        }
    });
    // 异步 httpclient 连接数配置
    builder.setHttpClientConfigCallback(
            new RestClientBuilder.HttpClientConfigCallback() {
```

```
            @Override
            public HttpAsyncClientBuilder customizeHttpClient(
                    HttpAsyncClientBuilder httpClientBuilder) {
                httpClientBuilder.setMaxConnTotal(maxConnectNum);
                httpClientBuilder.setMaxConnPerRoute(maxConnectPerRoute);
                return httpClientBuilder;
            }
        });

        RestHighLevelClient client = new RestHighLevelClient(builder);
        return client;
    }
}
```

5. 创建数据类

从网络获取数据,并且将数据导入到 Elasticsearch 中,获取网络数据,并解析网页的数据,实例代码如下。

```
package com.it.jareny.bigdata.elasticsearch.entity;

import lombok.AllArgsConstructor;
import lombok.Builder;
import lombok.Data;
import lombok.NoArgsConstructor;

@Data
@AllArgsConstructor
@NoArgsConstructor
@Builder
public class Content {
    private String title;
    private String img;
    private String price;
}
```

网页数据解析,实例代码如下。

```
package com.it.jareny.bigdata.elasticsearch.utils;

import com.it.jareny.bigdata.elasticsearch.entity.Content;
import org.jsoup.Jsoup;
import org.jsoup.nodes.Document;
import org.jsoup.nodes.Element;
import org.jsoup.select.Elements;
import org.springframework.stereotype.Component;
import java.net.URL;
import java.util.ArrayList;
```

```java
import java.util.List;

@Component
public class HtmlParseUtil {
    public List<Content> parseJD(String keywords) throws Exception {
        // 搜索
        String url = "https://search.jd.com/Search? keyword="+keywords;
        // 解析网页
        Document document = Jsoup.parse(new URL(url), 30000);
        Element element =document.getElementById("J_goodsList");
        // 获取所有的 el 标签
        Elements elements = element.getElementsByTag("li");
        List<Content> goodsList = new ArrayList<>();
        for (Element elet:elements){
            String img = elet.getElementsByTag("img")
                .eq(0).attr("data-lazy-img");
            String price = elet.getElementsByClass("p-price").eq(0).text();
            String title = elet.getElementsByClass("p-name").eq(0).text();
            Content content = Content.builder().img(img).price(price)
                .title(title).build();
            goodsList.add(content);
        }
        return goodsList;
    }
}
```

10.4.2 Elasticsearch 实现搜索功能

搜索功能的实现分两部分，分别为控制层和业务层，请求控制层接收用户发起的请求，业务层是逻辑处理层，用于处理用户发起的请求。

1. 控制层

项目的控制层，就是为了接收用户搜索的请求，将搜索的请求路径对外暴露给用户使用，然后用户输入搜索的关键词就可以搜索到相关的内容。

```java
package com.it.jareny.bigdata.elasticsearch.controller;

import com.it.jareny.bigdata.elasticsearch.service.impl.ContentService;
import org.springframework.beans.factory.annotation.Autowired;
import org.springframework.web.bind.annotation.GetMapping;
import org.springframework.web.bind.annotation.PathVariable;
import org.springframework.web.bind.annotation.ResponseBody;
import org.springframework.web.bind.annotation.RestController;

import java.util.List;
import java.util.Map;
```

```java
@RestController
@ResponseBody
public class ContentController {
    @Autowired
    private ContentService  contentService;

    @GetMapping("/parse/{keyword}")
    public Boolean parse(
            @PathVariable("keyword") String keyword) throws Exception {
        return  contentService.parseContent(keyword);
    }

    @GetMapping("/test")
    public String tesr() throws Exception {
        return  "test ok";
    }

    @GetMapping("/search/{keyword}/{pageNo}/{pageSize}")
    public List<Map<String, Object>> search(
            @PathVariable("keyword") String keyword,
            @PathVariable("pageNo") int pageNo,
            @PathVariable("pageSize") int pageSize) throws Exception {
        return contentService.searchPage(keyword,pageNo,pageSize);
    }

    @GetMapping("/searchPageHighlight/{keyword}/{pageNo}/{pageSize}")
    public List<Map<String, Object>> searchPageHighlight(
            @PathVariable("keyword") String keyword,
            @PathVariable("pageNo") int pageNo,
            @PathVariable("pageSize") int pageSize) throws Exception {
        return contentService.searchPageHighlight(keyword,pageNo,pageSize);
    }
}
```

2. 业务层

业务层就是将用户输入的关键词转换成 Elasticsearch 搜索系统能够识别的搜索语句，然后从搜索系统查询出用户想要搜索的内容，并且根据用户输入的关键词按评分来排序，最后将搜索的结果返回给用户。

```java
package com.it.jareny.bigdata.elasticsearch.service.impl;

import com.alibaba.fastjson.JSON;
import com.it.jareny.bigdata.elasticsearch.entity.Content;
import com.it.jareny.bigdata.elasticsearch.utils.HtmlParseUtil;
import org.elasticsearch.action.bulk.BulkRequest;
```

```java
import org.elasticsearch.action.bulk.BulkResponse;
import org.elasticsearch.action.index.IndexRequest;
import org.elasticsearch.action.search.SearchRequest;
import org.elasticsearch.action.search.SearchResponse;
import org.elasticsearch.client.RequestOptions;
import org.elasticsearch.client.RestHighLevelClient;
import org.elasticsearch.common.text.Text;
import org.elasticsearch.common.unit.TimeValue;
import org.elasticsearch.common.xcontent.XContentType;
import org.elasticsearch.index.query.TermQueryBuilder;
import org.elasticsearch.search.SearchHit;
import org.elasticsearch.search.builder.SearchSourceBuilder;
import org.elasticsearch.search.fetch.subphase.highlight.HighlightBuilder;
import org.elasticsearch.search.fetch.subphase.highlight.HighlightField;
import org.springframework.beans.factory.annotation.Autowired;
import org.springframework.beans.factory.annotation.Qualifier;
import org.springframework.stereotype.Service;

import java.io.IOException;
import java.util.ArrayList;
import java.util.List;
import java.util.Map;
import java.util.concurrent.TimeUnit;

@Service
public class ContentService {
    @Qualifier("elasticsearchClient")
    @Autowired
    private RestHighLevelClient   restHighLevelClient;

    // 1.把查询数据写入 ES
    public Boolean parseContent(String keywords) throws Exception {
        List<Content> contentList = new HtmlParseUtil().parseJD(keywords);
        BulkRequest bulkRequest = new BulkRequest();
        bulkRequest.timeout("2m");
        for(Content content:contentList){
            bulkRequest.add(new IndexRequest("jd_goods")
             .source(JSON.toJSONString(content),XContentType.JSON));
        }
        BulkResponse bulk = restHighLevelClient.bulk(
                bulkRequest,RequestOptions.DEFAULT);
        return  ! bulk.hasFailures();
    }

    // 2.获取数据,实现搜索功能
    public List<Map<String, Object>> searchPage(
```

```java
        String keyword, int pageNo, int pageSize) throws IOException {
    if(pageNo<=1){
        pageNo=1;
    }
    // 条件搜索
    SearchRequest searchRequest = new SearchRequest("jd_goods");
    SearchSourceBuilder sourceBuilder = new SearchSourceBuilder();

    // 分页
    sourceBuilder.from(pageNo);
    sourceBuilder.size(pageSize);

    // 精确匹配
    TermQueryBuilder termQueryBuilder =
            new TermQueryBuilder("title",keyword);
    sourceBuilder.query(termQueryBuilder);
    sourceBuilder.timeout(new TimeValue(60, TimeUnit.SECONDS));

    // 执行搜索
    searchRequest.source(sourceBuilder);
    SearchResponse searchResponse =
            restHighLevelClient.search(
    searchRequest,RequestOptions.DEFAULT);
    // 解析结果
    List<Map<String, Object>> list = new ArrayList<>();
    for(SearchHit documentFields:searchResponse.getHits().getHits()){
        Map<String, Object> sourceAsMap = documentFields.getSourceAsMap();
        list.add(sourceAsMap);
    }
    return list;
}

// 3.获取数据,实现搜索功能高亮显示
public List<Map<String, Object>> searchPageHighlight(
        String keyword, int pageNo, int pageSize) throws IOException {
    if(pageNo<=1){
        pageNo=1;
    }
    // 第一步条件搜索
    SearchRequest searchRequest = new SearchRequest("jd_goods");
    SearchSourceBuilder sourceBuilder = new SearchSourceBuilder();

    // 第二步分页
    sourceBuilder.from(pageNo);
    sourceBuilder.size(pageSize);
```

```java
    // 第三步精确匹配
    TermQueryBuilder termQueryBuilder =
            new TermQueryBuilder("title",keyword);
    sourceBuilder.query(termQueryBuilder);
    sourceBuilder.timeout(new TimeValue(60, TimeUnit.SECONDS));

    // 第四步高亮
    HighlightBuilder highlightBuilder = new HighlightBuilder();
    highlightBuilder.field("title");
    highlightBuilder.requireFieldMatch(true);
    highlightBuilder.preTags("<span style='color:red'>");
    highlightBuilder.postTags("</span>");
    sourceBuilder.highlighter(highlightBuilder);

    // 第五步执行搜索
    searchRequest.source(sourceBuilder);
    SearchResponse searchResponse =
            restHighLevelClient.search(
    searchRequest,RequestOptions.DEFAULT);
    // 第一步解析结果
    List<Map<String, Object>> list = new ArrayList<>();
    for(SearchHit documentFields:searchResponse.getHits().getHits()){
        Map<String, HighlightField> highlightFields =
            documentFields.getHighlightFields();
        HighlightField title = highlightFields.get("title");
        Map<String, Object> sourceAsMap = documentFields.getSourceAsMap();
        if(title!=null){
            Text[] fragments = title.fragments();
            String new_title ="";
            for(Text text:fragments){
                new_title += text;
            }
            sourceAsMap.put("title",new_title);
        }
        list.add(sourceAsMap);
    }
    return list;
    }
}
```

第11章 分布式处理引擎 Flink

本章内容介绍 Flink 的基本概念和工作原理,包括 DataStream 和 DataSet API、Flink 运行架构等。学习 Flink 的安装和配置,以及集群模式的部署。学习 Flink 的高级特性和应用场景,包括 Window、State、CEP 等功能。实践 Flink 的应用案例,包括流式处理、批处理等场景。

11.1 Flink 概述

1. Flink 简介

Flink 是一个开源的分布式、高性能、高可用、准确的流处理框架。支持实时流处理和批处理。Flink 支持多种数据源和数据接收器,包括 Kafka、HDFS、ElasticSearch 等。

2. Flink 特点

Flink 的核心特点,包括实时流处理、批处理和状态计算。Flink 提供了实时流处理的能力,可以处理实时数据流,并保证延迟和吞吐量的高效性。Flink 特点如下。

1)高吞吐量和低延迟:Flink 的流处理可以在毫秒级别内处理事件,同时保持高吞吐量。
2)容错性:Flink 提供了多种容错机制,检查点和故障恢复,以确保数据处理的可靠性。
3)分布式处理:Flink 可以在分布式环境中运行,可以处理大规模数据集。
4)灵活性:Flink 支持多种数据源和数据格式,并且可以与其他工具和框架集成。
5)状态管理:Flink 提供了状态管理机制,可以在流处理过程中跟踪和管理状态。
6)事件时间处理:Flink 支持事件时间处理,可以处理无序事件流,也可以处理延迟事件。
7)精确一次处理:Flink 支持精确一次处理,可以确保每个事件只被处理一次。

3. Flink 的应用场景

Flink 在各种实际业务场景中的应用。了解 Flink 在实时流处理、批处理和状态计算等方面的应用场景,并掌握如何将 Flink 应用于实际业务问题。

1)实时数据处理:Flink 可以处理实时数据流,例如从传感器、日志、社交媒体等收集的数据。它可以对数据进行过滤、转换、聚合等操作,并将结果输出到数据库、文件系统或其他数据存储系统中。

2）批处理：Flink 可以处理大规模的批处理作业，例如从 HDFS 中读取数据，对数据进行处理，并将结果输出到 HDFS 或其他数据存储系统中。

3）机器学习：Flink 可以用于机器学习任务，例如分类、聚类、回归等。它可以处理大规模的数据集，并使用分布式算法进行计算。

4）实时推荐：Flink 可以用于实时推荐系统，例如根据用户的行为实时推荐商品或内容。它可以处理实时数据流，并使用机器学习算法进行推荐。

5）金融风控：Flink 可以用于金融风控任务，例如实时监测交易数据、检测欺诈行为。它可以处理实时数据流，并使用机器学习算法进行分析。

11.2 Flink 基本组件和运行时架构

本节将探讨 Flink 的基本组件和运行时架构，以及如何利用这些组件和架构来构建高效的流处理应用。了解 Flink 的基本组件、运行时架构、数据处理方式和工作原理，并掌握如何将其应用于实际业务场景。

11.2.1 Flink 运行时架构

了解 Flink 运行时架构的各个组件和工作原理，帮助您更好地理解和应用 Flink。Flink 运行时架构主要由以下几个组件组成，如图 11-1 所示。

Flink 运行时架构主要包括作业管理器（Job Manager）、资源管理器（Resource Manager）、任务管理器（Task Manager）以及分发器（Dispatcher）四个组件。它们会在运行流处理应用程序时协同工作，如下。

（1）作业管理器

控制一个应用程序执行的主进程，也就是说，每个应用程序都会被一个不同的作业管理器所控制执行。

作业管理器会先接收到要执行的应用程序，这个应用程序会包括：作业图（Job graph），逻辑数据流图（Logical dataflow graph），以及打包了所有的类、库和其他资源的 JAR 包。

● 图 11-1　Flink 运行时架构

作业管理器会把 Job graph 转换成一个物理层面的数据流图，这个图被叫作"执行图"（Execution graph），包含了所有可以并发执行的任务。作业管理器会向资源管理器请求执行任务必要的资源，也就是任务管理器上的插槽。一旦它获取到了足够的资源，就会将执行图分发到真正运行它们的任务管理器上。而在运行过程中作业管理器会负责所有需要中央协调的操作，比如说检查点的协调。

（2）任务管理器

Flink 中的工作进程。通常在 Flink 中会有多个任务管理器运行，每个任务管理器都包含了一定数量的插槽。插槽的数量限制了任务管理器能够执行的任务数量。

启动之后，任务管理器会向资源管理器注册它的插槽；收到资源管理器的指令后，任务管理器就会将一个或者多个插槽提供给作业管理器调用。作业管理器就可以向插槽分配任务来执行了。

在执行过程中，一个任务管理器可以跟其他运行同一应用程序的任务管理器交换数据。

（3）资源管理器

主要负责管理任务管理器的插槽。任务管理器插槽是 Flink 中定义的处理资源单元。

Flink 为不同的环境和资源管理工具提供了不同资源管理器，比如 YARNMesos、K8s，以及 standalone 部署。

当作业管理器申请插槽资源时，资源管理器会将有空闲插槽的任务管理器分配给作业管理器。如果资源管理器没有足够的插槽来满足作业管理器的请求，它还可以向资源提供平台发起会话，以提供启动任务管理器进程的容器。

（4）分发器

可以跨作业运行，它为应用提交提供了 REST 接口。当一个应用被提交执行时，分发器就会启动并将应用移交给作业管理器，然后分发器会启动一个 WebUi，用来方便地展示和监控作业执行的信息。

11.2.2 Flink 的分层

Flink 是一个分层架构的系统，每一层所包含的组件都提供了特定的抽象，用来服务于上层组件。Flink 分层的组件栈，如图 11-2 所示。

Flink 架构可以分为四层，包括部署层、核心层、接口层和扩展库层，它们的作用如下。

1）部署层：该层主要涉及 Flink 的部署模式，Flink 支持多种部署模式——本地、集和云服务器。

2）核心层：该层提供了支持 Flink 计算的全部核心实现，为 API 层提供基础服务。

3）接口层：该层主要实现了面向无界 Stream 的流处理和面向 Batch 的批处理 API，其中流处理对应 DataStream API，批处理对应 DataSet API。

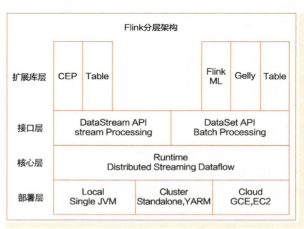

● 图 11-2　Flink 分层的组件栈

4）扩展库层：该层也被称为 Flink 应用框架层，根据 API 层的划分，在 API 层之上构建的满足特定应用的实现计算框架，也分别对应面向流处理和面向批处理两类。面向流处理支持 CEP（复杂事件处理）、基于 SQL-like 的操作（基于 Table 的关系操作）；面向批处理支持 FlinkML（机器学习库）、Gelly（图处理）、Table 操作。

11.3 Flink 流处理流程

Flink 流处理 API 主要分为四部分：Environment 环境、Source 数据源、Transform 转换、Sink 输出，如图 11-3 所示。

● 图 11-3　Flink 流处理

Flink 应用程序都需要一个执行环境，每一个组件的作用如下。

1）数据源：Flink 支持多种数据源，包括 Kafka、HDFS、Cassandra 等。在流处理开始之前，需要从数据源中读取数据，并将其转换为 Flink 内部的数据格式。

2）数据转换：在 Flink 中，数据以流的形式传输，每个算子都可以接收一个或多个输入流，并产生一个或多个输出流。为了对数据进行转换，需要编写算子，并将它们组合成一个流水线。在流水线中，每个算子都会接收上一个算子产生的输出流，并将其转换为下一个算子可以接收的输入流。

3）状态管理：在流处理过程中，可能需要对数据进行聚合、过滤、排序等操作，并且需要保存中间结果。为了实现这些操作，需要使用状态管理组件。在 Flink 中，状态可以存储在内存、文件系统或外部数据库中，以便在发生故障时进行恢复。

4）设置检查点（Checkpoint）：为了保证流处理的容错性，需要定期对状态进行备份。在 Flink 中，可以使用 Checkpoint 来实现状态备份。Checkpoint 会定期将作业的状态保存到持久化存储中，并在发生故障时进行恢复。

5）数据接收器：在流处理结束之后，需要将处理结果写入数据接收器，如 Kafka、HDFS、Cassandra 等。在 Flink 中，可以使用数据接收器来实现这些操作。

总之，Flink 流处理的流程包括数据源、数据转换、状态管理、Checkpoint 和数据接收器等组件。通过这些组件的协作，可以实现高吞吐量、低延迟、高可用性和高容错性的流处理。

11.3.1 Flink 环境设置（Environment）

Flink 程序可以在各种上下文环境中运行，提交作业执行计算时，首先必须获取当前 Flink 的运行环境，从而建立起与 Flink 框架之间的联系。只有获取了环境上下文信息，才能将具体的任务调度到不同的任务管理执行。

在获取到程序执行环境后，我们还可以对执行环境进行灵活的设置。比如可以全局设置程序的并行度、禁用算子链，还可以定义程序的时间语义、配置容错机制。

1. Flink 支持环境

了解 Flink 支持环境的基本要素、配置方法和优化技巧,并掌握如何为 Flink 应用提供最佳的运行时环境。Flink 支持环境如下。

1)LocalEnvironment 本地模式执行。
2)RemoteEnvironment 提交到远程集群执行。
3)CollectionEnvironment 集合数据集模式执行。
4)OptimizerPlanEnvironment 不执行作业,仅创建优化的计划。
5)PreviewPlanEnvironment 提取预先优化的执行计划。
6)ContextEnvironment 用于在客户端上远程执行。
7)DetachedEnvironment 用于在客户端上以分离模式进行远程执行。

2. 获取执行环境

创建一个执行环境,表示当前执行程序的上下文。要在 Java 中获取执行环境,可以使用 Flink 提供的 ExecutionEnvironment.getExecutionEnvironment() 方法,它会根据查询运行的方式决定返回什么样的运行环境。

```java
import org.apache.flink.api.java.ExecutionEnvironment;
public class GetExecutionEnvironment {
    public static void main(String[] args) throws Exception {
        ExecutionEnvironment env = ExecutionEnvironment.getExecutionEnvironment();
        System.out.println("Execution environment: " + env);
    }
}
```

3. 创建本地执行环境

在 Flink 中创建本地执行环境,可以使用 ExecutionEnvironment.createLocalEnvironment() 方法。该方法返回一个 ExecutionEnvironment 类的实例,表示 Flink 程序的本地执行环境。需要在调用时指定默认的并行度。如果不传入,则默认并行度就是本地的 CPU 核心数。

```java
import org.apache.flink.api.java.ExecutionEnvironment;
public class CreateLocalEnvironment {
    public static void main(String[] args) throws Exception {
        ExecutionEnvironment env = ExecutionEnvironment.createLocalEnvironment();
        System.out.println("Local execution environment: " + env);
    }
}
```

4. 创建远程执行环境

在 Flink 中创建远程执行环境,可以使用 ExecutionEnvironment.createRemoteEnvironment() 方法。该方法需要传入 Flink 集群的主机名和端口号,并返回一个 ExecutionEnvironment 类的实例,表示 Flink 程序的远程执行环境。

```
import org.apache.flink.api.java.ExecutionEnvironment;
public class CreateRemoteEnvironment {
    public static void main(String[] args) throws Exception {
        ExecutionEnvironment env = ExecutionEnvironment.createRemoteEnvironment("local-
host"; 6123);
        System.out.println("Remote execution environment: " + env);
    }
}
```

5. 执行模式

（1）流执行模式（STREAMING）

Flink 流执行模式是一种用于处理无限数据流的执行模式。这是 DataStream API 最经典的模式，一般用于需要持续实时处理的无界数据流。

以下是一个使用 Java 编写的 Flink 流执行模式的示例代码：

```
import org.apache.flink.streaming.api.environment.StreamExecutionEnvironment;

public class StreamingJob {
    public static void main(String[] args) throws Exception {
        StreamExecutionEnvironment env = StreamExecutionEnvironment.getExecutionEnvironment();
        env.socketTextStream("localhost", 9999)
            .flatMap((String line, Collector<String> out) -> {
                for (String word : line.split(" ")) {
                    out.collect(word);
                }
            })
            .keyBy(word -> word)
            .sum(1)
            .print();
        env.execute("Streaming WordCount");
    }
}
```

（2）批执行模式（BATCH）

Flink 批执行模式是一种用于处理有限数据集的执行模式。在 Flink 中，可以使用 ExecutionEnvironment 类来创建批执行环境，并实现批处理任务。

以下是一个使用 Java 编写的 Flink 批执行模式的示例代码：

```
import org.apache.flink.api.java.ExecutionEnvironment;
import org.apache.flink.api.java.DataSet;

public class BatchJob {
    public static void main(String[] args) throws Exception {
        ExecutionEnvironment env = ExecutionEnvironment.getExecutionEnvironment();
        // 设置批执行模式
```

```
        env.setRuntimeMode(RuntimeExecutionMode.BATCH);
        DataSet<String> text = env.readTextFile("path/to/input/file");
        DataSet<String> words = text.flatMap((String line, Collector<String> out) -> {
            for (String word : line.split(" ")) {
                out.collect(word);
            }
        });
        DataSet<Tuple2<String, Integer>> counts = words.groupBy(0).sum(1);
        counts.writeAsCsv("path/to/output/file");
        env.execute("Batch WordCount");
    }
}
```

（3）自动模式（AUTOMATIC）

Flink 自动执行模式是一种自动化的执行模式，它可以根据数据流的特性自动选择批处理或流处理模式。在 Flink 中，可以使用 StreamExecutionEnvironment 类来创建流执行环境，使用 ExecutionEnvironment 类来创建批执行环境。Flink 自动执行模式可以根据输入数据的特性自动选择最佳的执行模式，从而实现更高效的数据处理。

6. 触发程序执行

Flink 是由事件驱动的，只有等到数据到来，才会触发真正的计算，这也被称为"延迟执行"或"懒执行"（lazy execution）。所以我们需要显式地调用执行环境的 execute 方法，来触发程序执行。execute 方法将一直等待作业完成，然后返回一个执行结果。

▶▶ 11.3.2　Flink 源算子（Source）

Flink 源算子是一种用于从外部数据源读取数据的算子。Flink 支持多种数据源，包括 Kafka、HDFS、Cassandra 等。在 Flink 中，可以使用 StreamExecutionEnvironment.addSource()方法或 ExecutionEnvironment.readTextFile()方法来创建源算子。

以下是一个使用 Java 编写的 Flink 源算子的示例代码。

```java
import org.apache.flink.streaming.api.environment.StreamExecutionEnvironment;
import org.apache.flink.streaming.api.functions.source.SourceFunction;

public class SourceJob {
    public static void main(String[] args) throws Exception {
        StreamExecutionEnvironment env = StreamExecutionEnvironment.getExecutionEnvironment();
        env.addSource(new SourceFunction<String>() {
            @Override
            public void run(SourceContext<String> ctx) throws Exception {
                while (true) {
                    ctx.collect("Hello, world!");
                    Thread.sleep(1000);
                }
            }
```

```
            @Override
            public void cancel() {}
    }).print();
    env.execute("Source Job");
    }
}
```

11.3.3 Flink 支持的数据类型

Flink 可以从各种来源获取数据，然后构建 DataStream 进行转换处理。一般将数据的输入来源称为数据源（data source），而读取数据的算子就是源算子（source operator）。所以，source 就是我们整个处理程序的输入端。

1. Flink 支持的数据类型

1）基本数据类型，如 int、long、float、double、boolean 等。
2）复合数据类型，如 Tuple、Case Class、POJO 等。
3）Flink 提供的特殊数据类型，如 Row、Value、List、Map 等。
4）用户自定义数据类型，可以通过实现 TypeInformation 接口来定义自己的数据类型。

2. Flink 读取数据的案例

1）以下是使用 Flink 从集合中读取数据的示例代码。

```
import org.apache.flink.api.java.ExecutionEnvironment;
import org.apache.flink.api.java.DataSet;

public class CollectionJob {
    public static void main(String[] args) throws Exception {
        ExecutionEnvironment env = ExecutionEnvironment.getExecutionEnvironment();
        DataSet<String> data = env.fromElements("Hello", "World", "Flink");
        data.print();
    }
}
```

2）以下是使用 Flink 从文件中读取数据的示例代码。

```
import org.apache.flink.api.java.ExecutionEnvironment;
import org.apache.flink.api.java.DataSet;

public class ReadFileJob {
    public static void main(String[] args) throws Exception {
        ExecutionEnvironment env = ExecutionEnvironment.getExecutionEnvironment();
        DataSet<String> data = env.readTextFile("path/to/file.txt");
        data.print();
    }
}
```

3）以下是使用 Flink 从 Socket 中读取数据的示例代码。

```java
import org.apache.flink.streaming.api.environment.StreamExecutionEnvironment;
import org.apache.flink.streaming.api.datastream.DataStream;
import org.apache.flink.streaming.api.functions.source.SocketTextStreamFunction;

public class ReadSocketJob {
    public static void main(String[] args) throws Exception {
        StreamExecutionEnvironment env = StreamExecutionEnvironment.getExecutionEnvironment();
        DataStream<String> data = env.addSource(new SocketTextStreamFunction("localhost", 9999, "\n", 3));
        data.print();
        env.execute("Read from Socket");
    }
}
```

4）以下是使用 Flink 从 Kafka 中读取数据的示例代码。

```java
import org.apache.flink.streaming.api.environment.StreamExecutionEnvironment;
import org.apache.flink.streaming.api.datastream.DataStream;
import org.apache.flink.streaming.connectors.kafka.FlinkKafkaConsumer;
import java.util.Properties;

public class ReadKafkaJob {
    public static void main(String[] args) throws Exception {
        StreamExecutionEnvironment env = StreamExecutionEnvironment.getExecutionEnvironment();
        Properties properties = new Properties();
        properties.setProperty("bootstrap.servers", "localhost:9092");
        properties.setProperty("group.id", "test");
        FlinkKafkaConsumer<String> consumer = new FlinkKafkaConsumer<>("my-topic", new SimpleStringSchema(), properties);
        DataStream<String> data = env.addSource(consumer);
        data.print();
        env.execute("Read from Kafka");
    }
}
```

11.3.4　Flink 转换算子（Transform）

Flink 转换算子是 Flink 数据流处理中的核心。它们是用于将一个或多个数据流转换为另一个数据流的函数。Flink 提供了许多内置的转换算子，例如 map、filter、flatMap、keyBy、reduce、window 等。这些算子可以被组合在一起，以构建复杂的数据流处理应用程序。此外，Flink 还允许用户编写自定义的转换算子，以满足特定的需求。转换算子可以应用于批处理和流处理场景，但在流处理场景中，它们通常是无限的，即可以处理无限的数据流。

11.3.5　Flink 输出算子（Sink）

Flink 的 DataStream API 专门提供了向外部写入数据的方法 addSink，主要就是用来实现与外部系统连接，并将数据提交写入的 Sink；Flink 程序中所有对外的输出操作，一般都是利用 Sink 算子完成的。

Flink 输出数据的案例：

1）以下是使用 Flink 将数据流中的元素写入指定文本文件中的代码。

```java
import org.apache.flink.api.common.functions.FlatMapFunction;
import org.apache.flink.api.java.utils.ParameterTool;
import org.apache.flink.streaming.api.datastream.DataStream;
import org.apache.flink.streaming.api.environment.StreamExecutionEnvironment;
import org.apache.flink.util.Collector;

public class WriteToFile {
    public static void main(String[] args) throws Exception {
        // 从命令行参数中读取文件路径
        final ParameterTool params = ParameterTool.fromArgs(args);
        final String outputPath = params.get("output");
        // 获取执行环境
        final StreamExecutionEnvironment env = StreamExecutionEnvironment.getExecutionEnvironment();
        // 读取数据流
        DataStream<String> input = env.socketTextStream("localhost", 9999);
        // 将数据流中的元素写入指定的文本文件中
        input.writeAsText(outputPath);
        // 执行任务
        env.execute("Write to File");
    }
}
```

2）输出到 Kafka。Flink 从 Kakfa 的一个 topic 读取消费数据，然后进行处理转换，最终将结果数据写入 Kafka 的另一个 topic。数据从 Kafka 流入，经 Flink 处理后又流回 Kafka，这就是所谓的数据管道应用。

Flink 与 Kafka 的连接器提供了端到端的精确一次语义保证，这在实际项目中是最高级别的一致性保证。

以下是使用 Flink 将数据流中的元素写入指定 Kafka 主题中的代码。

```java
import org.apache.flink.api.common.serialization.SimpleStringSchema;
import org.apache.flink.api.java.utils.ParameterTool;
import org.apache.flink.streaming.api.datastream.DataStream;
import org.apache.flink.streaming.api.environment.StreamExecutionEnvironment;
import org.apache.flink.streaming.connectors.kafka.FlinkKafkaProducer;
import org.apache.flink.streaming.connectors.kafka.KafkaSerializationSchema;
import org.apache.kafka.clients.producer.ProducerConfig;
import org.apache.kafka.clients.producer.ProducerRecord;
import org.apache.kafka.common.serialization.StringSerializer;
```

```java
import java.util.Properties;

public class WriteToKafka {
    public static void main(String[] args) throws Exception {
        // 从命令行参数中读取 Kafka 相关配置
        final ParameterTool params = ParameterTool.fromArgs(args);
        final String bootstrapServers = params.get("bootstrap.servers");
        final String topic = params.get("topic");
        // 获取执行环境
        final StreamExecutionEnvironment env = StreamExecutionEnvironment.getExecutionEnvironment();
        // 读取数据流
        DataStream<String> input = env.socketTextStream("localhost", 9999);
        // 将数据流中的元素写入指定的 Kafka 主题中
        Properties props = new Properties();
        props.setProperty(ProducerConfig.BOOTSTRAP_SERVERS_CONFIG, bootstrapServers);
        FlinkKafkaProducer<String> producer = new FlinkKafkaProducer<>(topic, new SimpleStringSchema(), props);
        input.addSink(producer);
        // 执行任务
        env.execute("Write to Kafka");
    }
}
```

3）以下是使用 Flink 将数据流中的数据写入 Redis 中的代码。

```java
import org.apache.flink.api.common.functions.FlatMapFunction;
import org.apache.flink.api.java.utils.ParameterTool;
import org.apache.flink.streaming.api.datastream.DataStream;
import org.apache.flink.streaming.api.environment.StreamExecutionEnvironment;
import org.apache.flink.streaming.connectors.redis.RedisSink;
import org.apache.flink.streaming.connectors.redis.common.config.FlinkJedisConfigBase;
import org.apache.flink.streaming.connectors.redis.common.config.FlinkJedisPoolConfig;
import org.apache.flink.streaming.connectors.redis.common.mapper.RedisCommand;
import org.apache.flink.streaming.connectors.redis.common.mapper.RedisCommandDescription;
import org.apache.flink.streaming.connectors.redis.common.mapper.RedisMapper;
import org.apache.flink.util.Collector;

public class WriteToRedis {
    public static void main(String[] args) throws Exception {
        // 从命令行参数中读取 Redis 相关配置
        final ParameterTool params = ParameterTool.fromArgs(args);
        final String redisHost = params.get("redis.host");
        final int redisPort = params.getInt("redis.port");
        final String redisKey = params.get("redis.key");
```

```
    // 获取执行环境
    final StreamExecutionEnvironment env = StreamExecutionEnvironment.getExecutionEnvironment();

    // 读取数据流
    DataStream<String> input = env.socketTextStream("localhost", 9999);

    // 将数据流中的元素写入指定的 Redis 中
    FlinkJedisPoolConfig jedisPoolConfig = new FlinkJedisPoolConfig.Builder()
            .setHost(redisHost)
            .setPort(redisPort)
            .build();
    RedisSink<String> redisSink = new RedisSink<>(jedisPoolConfig, new RedisMapper<String>() {
        @Override
        public RedisCommandDescription getCommandDescription() {
            return new RedisCommandDescription(RedisCommand.SET);
        }

        @Override
        public String getKeyFromData(String data) {
            return redisKey;
        }

        @Override
        public String getValueFromData(String data) {
            return data;
        }
    });
    input.addSink(redisSink);

    // 执行任务
    env.execute("Write to Redis");
}
```

4) 以下是使用 Flink 基于数据流采集数据写入 Elasticsearch 的代码示例。

```
import org.apache.flink.streaming.api.environment.StreamExecutionEnvironment;
import org.apache.flink.streaming.connectors.elasticsearch.ElasticsearchSink;
import org.apache.flink.streaming.connectors.elasticsearch6.ElasticsearchSinkBuilder;
import org.apache.flink.api.common.functions.RuntimeContext;
import org.apache.flink.api.common.serialization.SimpleStringSchema;
import org.apache.flink.api.common.typeinfo.TypeInformation;
import org.apache.flink.api.java.utils.ParameterTool;
import org.apache.flink.streaming.api.functions.sink.SinkFunction;
import org.apache.flink.streaming.connectors.elasticsearch.RequestIndexer;
```

```java
import org.apache.flink.streaming.connectors.elasticsearch.util.ElasticsearchSinkUtil;
import org.apache.flink.streaming.connectors.elasticsearch.util.FailureHandler;
import org.elasticsearch.action.index.IndexRequest;
import org.elasticsearch.client.Requests;
import org.elasticsearch.common.xcontent.XContentType;

import java.util.ArrayList;
import java.util.HashMap;
import java.util.List;
import java.util.Map;

public class FlinkElasticsearchSink {
    public static void main(String[] args) throws Exception {
        final ParameterTool params = ParameterTool.fromArgs(args);
        final String hostname = params.get("hostname", "localhost");
        final int port = params.getInt("port", 9200);
        final String index = params.get("index", "flink-elasticsearch-sink");
        final String type_name = params.get("type", "flink-elasticsearch-sink-type");

        final StreamExecutionEnvironment env = StreamExecutionEnvironment.getExecutionEnvironment();

        List<Map<String, String>> dataList = new ArrayList<>();
        Map<String, String> data = new HashMap<>();
        data.put("name", "John");
        data.put("age", "30");
        dataList.add(data);

        ElasticsearchSink.Builder<Map<String, String>> esSinkBuilder = new ElasticsearchSink.Builder<>(
            ElasticsearchSinkUtil.createSinkFunction(hostname, port, index, type_name, new SimpleStringSchema(), new FailureHandler() {
                @Override
                public void onFailure(Throwable throwable, RequestIndexer requestIndexer) throws Throwable {
                    // handle failure
                }
            })
        );

        env.fromCollection(dataList)
            .map(dataMap -> {
                return dataMap.toString();
            })
            .addSink(esSinkBuilder.build());
```

```
            env.execute("Flink Elasticsearch Sink");
    }
}
```

5）以下是使用 Flink 将数据流中的元素写入 MySQL 的代码。

```
import org.apache.flink.api.common.functions.FlatMapFunction;
import org.apache.flink.api.java.utils.ParameterTool;
import org.apache.flink.streaming.api.datastream.DataStream;
import org.apache.flink.streaming.api.environment.StreamExecutionEnvironment;
import org.apache.flink.streaming.connectors.jdbc.JdbcConnectionOptions;
import org.apache.flink.streaming.connectors.jdbc.JdbcSink;
import org.apache.flink.streaming.connectors.jdbc.JdbcStatementBuilder;
import org.apache.flink.types.Row;
import java.sql.PreparedStatement;
import java.sql.SQLException;

public class WriteToMySQL {
    public static void main(String[] args) throws Exception {
        // 从命令行参数中读取 MySQL 相关配置
        final ParameterTool params = ParameterTool.fromArgs(args);
        final String mysqlUrl = params.get("mysql.url");
        final String mysqlUser = params.get("mysql.user");
        final String mysqlPassword = params.get("mysql.password");
        final String mysqlTable = params.get("mysql.table");
        // 获取执行环境
        final StreamExecutionEnvironment env = StreamExecutionEnvironment.getExecutionEnvironment();

        // 读取数据流
        DataStream<String> input = env.socketTextStream("localhost", 9999);

        // 将数据流中的元素写入指定的 MySQL 中
        JdbcConnectionOptions connectionOptions = new JdbcConnectionOptions.JdbcConnectionOptionsBuilder()
                .withUrl(mysqlUrl)
                .withUsername(mysqlUser)
                .withPassword(mysqlPassword)
                .build();
        JdbcStatementBuilder<Row> statementBuilder = new JdbcStatementBuilder<Row>() {
            @Override
            public void accept(PreparedStatement preparedStatement, Row row) throws SQLException {
                preparedStatement.setString(1, row.getField(0).toString());
            }
        };
        JdbcSink.sink(
            input.map(new FlatMapFunction<String, Row>() {
```

```
                @Override
                public void flatMap(String value, Collector<Row> out) throws Exception {
                    out.collect(Row.of(value));
                }
            }),
            JdbcSink.insertInto(mysqlTable).build(),
            statementBuilder,
            connectionOptions
        );
        // 执行任务
        env.execute("Write to MySQL");
    }
}
```

11.4 Flink 窗口、时间和水位线

11.4.1 Flink 窗口（Window）

Flink 中的窗口（Window）是一种将无限数据流划分为有限大小块的机制。窗口可以根据时间或其他条件进行划分，以便对数据进行有限的处理。Flink 中有两种类型的窗口：时间窗口和计数窗口。

时间窗口是基于时间的窗口，可以根据事件时间或处理时间进行划分。时间窗口可以根据窗口大小和滑动间隔进行划分。例如，一个 10 秒的时间窗口可以每 5 秒滑动一次，以便对数据进行处理。

计数窗口是基于事件数量的窗口，可以根据事件数量进行划分。计数窗口可以根据窗口大小和滑动间隔进行划分。例如，一个 100 个事件的计数窗口可以每 50 个事件滑动一次，以便对数据进行处理。

Flink 中的窗口可以用于各种数据处理任务，例如聚合、过滤和转换。窗口操作可以通过 WindowFunction 和 ProcessWindowFunction 来实现。WindowFunction 可以对窗口中的元素进行聚合操作，而 ProcessWindowFunction 可以对窗口中的元素进行更复杂的操作，例如连接外部系统或更新状态。

使用 Java 窗口，您需要使用 Flink 的 DataStream API。DataStream API 允许您以流的形式处理数据，并在处理数据时使用时间戳和水印来处理事件时间。下面是使用 Flink 自定义窗口编写的代码示例。

```
import org.apache.flink.api.common.functions.AggregateFunction;
import org.apache.flink.api.common.functions.ReduceFunction;
import org.apache.flink.api.java.tuple.Tuple;
import org.apache.flink.api.java.tuple.Tuple2;
import org.apache.flink.streaming.api.datastream.DataStream;
import org.apache.flink.streaming.api.environment.StreamExecutionEnvironment;
import org.apache.flink.streaming.api.functions.windowing.WindowFunction;
import org.apache.flink.streaming.api.windowing.assigners.TumblingEventTimeWindows;
import org.apache.flink.streaming.api.windowing.time.Time;
import org.apache.flink.streaming.api.windowing.windows.TimeWindow;
```

```java
import org.apache.flink.util.Collector;

public class CustomWindow {
    public static void main(String[] args) throws Exception {
        final StreamExecutionEnvironment env = StreamExecutionEnvironment.getExecutionEnvironment();

        DataStream<Tuple2<String, Integer>> stream = env.socketTextStream("localhost", 9999)
            .map(line -> {
                String[] tokens = line.split(",");
                return new Tuple2<>(tokens[0], Integer.parseInt(tokens[1]));
            });

        stream.keyBy(0)
            .window(TumblingEventTimeWindows.of(Time.seconds(10)))
            .aggregate(new AggregateFunction<Tuple2<String, Integer>, Integer, Integer>() {
                @Override
                public Integer createAccumulator() {
                    return 0;
                }

                @Override
                public Integer add(Tuple2<String, Integer> value, Integer accumulator) {
                    return accumulator + value.f1;
                }

                @Override
                public Integer getResult(Integer accumulator) {
                    return accumulator;
                }

                @Override
                public Integer merge(Integer a, Integer b) {
                    return a + b;
                }
            }, new WindowFunction<Integer, Tuple2<String, Integer>, Tuple, TimeWindow>() {
                @Override
                public void apply(Tuple key, TimeWindow window, Iterable<Integer> input, Collector<Tuple2<String, Integer>> out) throws Exception {
                    Integer sum = input.iterator().next();
                    out.collect(new Tuple2<>(key.getField(0), sum));
                }
            })
            .print();
```

```
        env.execute("Custom Window");
    }
}
```

11.4.2 Flink 时间（Time）

Flink 中有三种时间概念：事件时间（Event Time）、处理时间（Processing Time）和摄取时间（Ingestion Time）。

事件时间是事件实际发生的时间，通常由事件本身携带。事件时间是基于事件本身的，因此可以准确地反映事件的真实时间戳。在 Flink 中，事件时间是用来处理乱序事件的，可以通过事件时间来实现窗口操作和水印生成。

处理时间是 Flink 处理事件的本地系统时间。处理时间是基于 Flink 处理事件的时间，因此可以快速地处理事件。在 Flink 中，处理时间是默认的时间概念，可以通过 ProcessingTimeWindowAssigner 来实现基于处理时间的窗口操作。

摄取时间是事件进入 Flink 系统的时间。摄取时间是基于 Flink 接收事件的时间，因此可以反映事件进入 Flink 系统的时间。在 Flink 中，摄取时间可以通过 AssignerWithPeriodicWatermarks 来实现基于摄取时间的窗口操作。

使用 Flink 时间，您需要使用 Flink 的 DataStream API。DataStream API 允许您以流的形式处理数据，并在处理数据时使用时间戳和水印来处理事件时间。以下是 Flink 从 Kafka 主题中读取数据并使用事件时间对数据进行分组和聚合的代码。

```java
import org.apache.flink.api.common.functions.MapFunction;
import org.apache.flink.api.common.serialization.SimpleStringSchema;
import org.apache.flink.streaming.api.TimeCharacteristic;
import org.apache.flink.streaming.api.datastream.DataStream;
import org.apache.flink.streaming.api.environment.StreamExecutionEnvironment;
import org.apache.flink.streaming.connectors.kafka.FlinkKafkaConsumer;

import java.util.Properties;

public class FlinkTimeExample {
    public static void main(String[] args) throws Exception {
        // 创建执行环境
        StreamExecutionEnvironment env = StreamExecutionEnvironment.getExecutionEnvironment();

        // 设置时间特征为事件时间
        env.setStreamTimeCharacteristic(TimeCharacteristic.EventTime);

        // 配置 Kafka 消费者
        Properties props = new Properties();
        props.setProperty("bootstrap.servers", "localhost:9092");
        props.setProperty("group.id", "test");
```

```java
    // 创建 Kafka 消费者
    FlinkKafkaConsumer<String> consumer = new FlinkKafkaConsumer<>("my-topic", new SimpleStringSchema(), props);

    // 从最早的数据开始消费
    consumer.setStartFromEarliest();

    // 从 Kafka 主题中读取数据
    DataStream<String> stream = env.addSource(consumer);

    // 将数据转换为事件时间,并提取时间戳
    DataStream<MyEvent> events = stream.map(new MapFunction<String, MyEvent>() {
        @Override
        public MyEvent map(String value) throws Exception {
            // 解析数据并提取时间戳
            long timestamp = ...;
            return new MyEvent(timestamp, value);
        }
    });

    // 按事件时间分组,并进行聚合
    DataStream<MyAggregate> result = events
            .keyBy(event -> event.getTimestamp())
            .timeWindow(Time.minutes(1))
            .aggregate(new MyAggregator());

    // 输出结果
    result.print();

    // 执行任务
    env.execute("Flink Time Example");
}

// 定义事件类
public static class MyEvent {
    private long timestamp;
    private String value;

    public MyEvent(long timestamp, String value) {
        this.timestamp = timestamp;
        this.value = value;
    }

    public long getTimestamp() {
        return timestamp;
```

```java
        }

        public String getValue() {
            return value;
        }
    }

    // 定义聚合器
    public static class MyAggregator implements AggregateFunction<MyEvent, MyAggregate, MyAggregate> {
        @Override
        public MyAggregate createAccumulator() {
            return new MyAggregate();
        }

        @Override
        public MyAggregate add(MyEvent value, MyAggregate accumulator) {
            accumulator.addValue(value.getValue());
            return accumulator;
        }

        @Override
        public MyAggregate getResult(MyAggregate accumulator) {
            return accumulator;
        }

        @Override
        public MyAggregate merge(MyAggregate a, MyAggregate b) {
            a.merge(b);
            return a;
        }
    }

    // 定义聚合结果类
    public static class MyAggregate {
        private List<String> values = new ArrayList<>();

        public void addValue(String value) {
            values.add(value);
        }

        public void merge(MyAggregate other) {
            values.addAll(other.values);
        }

        public List<String> getValues() {
```

```
            return values;
        }
    }
}
```

在这个示例中，我们首先创建了一个执行环境，并将时间特征设置为事件时间。然后，配置了一个 Kafka 消费者，从 Kafka 主题中读取数据。接下来，将数据转换为事件时间，并按事件时间分组和聚合。最后，输出结果并执行任务。

▶▶ 11.4.3　Flink 水位线（Watermark）

Flink 水位线是一种机制，用于在事件时间处理中确定事件流的进度。在事件时间处理中，事件的时间戳是非递减的，但是事件的到达顺序是不确定的。因此，Flink 需要一种机制来确定事件流的进度，以便在处理时间窗口时，只考虑已经到达的事件。

水位线是一种特殊的事件，它表示事件时间流已经到达了某个时间点，且在此时间点之前的所有事件已经到达。因此，水位线可以用来触发时间窗口的计算。

11.5　Flink 状态管理

▶▶ 11.5.1　Flink 状态分类

Flink 中的状态可以分为两种类型：键控状态（Keyed State）和操作符状态（Operator State）。

键控状态是与特定键相关联的状态，通常用于在流处理中跟踪聚合值或其他与键相关的信息。键控状态可以通过 KeyedStateStore 来访问，可以在 ProcessFunction、RichFunction 和其他支持状态的函数中使用。

操作符状态是与操作符相关联的状态，通常用于在流处理中跟踪全局信息或其他与操作符相关的信息。操作符状态可以通过 OperatorStateStore 来访问，可以在 ProcessFunction、RichFunction 和其他支持状态的函数中使用。

Flink 中的状态可以通过 StateDescriptor 来定义，可以指定状态的名称、类型和默认值等属性。状态可以在运行时动态创建和访问，可以通过 StateTtlConfig 来设置状态的过期时间和清理策略。

▶▶ 11.5.2　Flink 状态后端

Flink 状态后端是指 Flink 用于存储和管理状态的后端系统。Flink 支持多种状态后端，包括内存、文件系统和分布式存储系统等。其中，内存状态后端是默认的状态后端，可以在本地内存中存储状态。文件系统状态后端可以将状态存储在本地文件系统中，而分布式存储系统状态后端可以将状态存储在分布式存储系统中，例如 HDFS、S3 或 RocksDB 等。Flink 中的状态后端可以通过 StateBackend 来配置，可以在作业级别或全局级别进行配置。在作业级别配置状态后端时，可以通过 ExecutionConfig 来设置状态后端的类型和相关参数。在全局级别配置状态后端时，可以通过 flink-conf.yaml 文件来设置状态后端的类型和相关参数。Flink 中的状态后端可以通过 Checkpoint 和 Savepoint 来实现状态的持久化和恢复，可以保证作业的容错性和可靠性。

11.5.3　Flink 状态管理案例

下面是一个使用 Java 编写的 Flink 状态管理的代码示例。

```java
import org.apache.flink.api.common.functions.MapFunction;
import org.apache.flink.api.common.state.ValueState;
import org.apache.flink.api.common.state.ValueStateDescriptor;
import org.apache.flink.api.java.tuple.Tuple;
import org.apache.flink.api.java.tuple.Tuple2;
import org.apache.flink.streaming.api.datastream.DataStream;
import org.apache.flink.streaming.api.environment.StreamExecutionEnvironment;
import org.apache.flink.streaming.api.functions.KeyedProcessFunction;
import org.apache.flink.util.Collector;

public class StateManagement {
    public static void main(String[] args) throws Exception {
        final StreamExecutionEnvironment env = StreamExecutionEnvironment.getExecutionEnvironment();

        DataStream<Tuple2<String, Integer>> stream = env.socketTextStream("localhost", 9999)
            .map(new MapFunction<String, Tuple2<String, Integer>>() {
                @Override
                public Tuple2<String, Integer> map(String line) throws Exception {
                    String[] tokens = line.split(",");
                    return new Tuple2<>(tokens[0], Integer.parseInt(tokens[1]));
                }
            });

        stream.keyBy(0)
            .process(new KeyedProcessFunction<Tuple, Tuple2<String, Integer>, Tuple2<String, Integer>>() {
                private transient ValueState<Integer> sumState;

                @Override
                public void open(org.apache.flink.configuration.Configuration parameters) throws Exception {
                    ValueStateDescriptor<Integer> descriptor = new ValueStateDescriptor<>("sum", Integer.class);
                    sumState = getRuntimeContext().getState(descriptor);
                }

                @Override
                public void processElement(Tuple2<String, Integer> value, Context ctx, Collector<Tuple2<String, Integer>> out) throws Exception {
```

```java
                Integer sum = sumState.value();
                if (sum == null) {
                    sum = 0;
                }
                sum += value.f1;
                sumState.update(sum);
                out.collect(new Tuple2<>(value.f0, sum));
            }
        })
        .print();
    env.execute("State Management");
    }
}
```

在这个示例中,我们首先创建了一个 StreamExecutionEnvironment 对象,然后使用 socketTextStream( ) 方法从本地的 9999 端口读取数据流。接下来,使用 keyBy( ) 方法将数据流按照第一个元素进行分组,然后使用 process( ) 方法定义了一个 KeyedProcessFunction,用于对每个分组内的数据进行状态管理。在 process( ) 方法中,我们首先使用 ValueStateDescriptor 创建了一个名为"sum"的状态变量,然后在 open( ) 方法中初始化了这个状态变量。在 processElement( ) 方法中,我们首先从状态变量中获取当前的状态值,然后将当前的数据值加入状态值中,并更新状态变量。最后,使用 Collector 将结果输出。

11.6 Flink 处理函数

11.6.1 Flink 处理函数

Flink 中的处理函数包括 ProcessFunction、KeyedProcessFunction、CoProcessFunction、KeyedCoProcessFunction、ProcessWindowFunction 和 KeyedProcessWindowFunction 等。

ProcessFunction 是 Flink 中最通用的处理函数,可以处理输入流中的每个元素,并生成零个、一个或多个输出元素。ProcessFunction 可以访问键控状态和操作符状态,并可以使用定时器来触发事件。ProcessFunction 可以通过实现 ProcessFunction 接口来定义,可以在 DataStream 上使用。

KeyedProcessFunction 是基于键控状态的 ProcessFunction,可以访问与特定键相关联的状态。KeyedProcessFunction 可以处理输入流中的每个元素,并生成零个、一个或多个输出元素。KeyedProcessFunction 可以使用定时器来触发事件,并可以访问键控状态和操作符状态。KeyedProcessFunction 可以通过实现 KeyedProcessFunction 接口来定义,可以在 KeyedStream 上使用。

CoProcessFunction 是用于连接两个流的处理函数,可以处理两个输入流中的每个元素,并生成零个、一个或多个输出元素。CoProcessFunction 可以访问键控状态和操作符状态,并可以使用定时器来触发事件。CoProcessFunction 可以通过实现 CoProcessFunction 接口来定义,可以在 ConnectedStreams 上使用。

KeyedCoProcessFunction 是基于键控状态的 CoProcessFunction，可以访问与特定键相关联的状态。KeyedCoProcessFunction 可以处理两个输入流中的每个元素，并生成零个、一个或多个输出元素。KeyedCoProcessFunction 可以使用定时器来触发事件，并可以访问键控状态和操作符状态。KeyedCoProcessFunction 可以通过实现 KeyedCoProcessFunction 接口来定义，可以在 ConnectedStreams 上使用。

ProcessWindowFunction 是用于窗口操作的处理函数。它可以对窗口中的元素进行聚合操作，并生成零个、一个或多个输出元素。ProcessWindowFunction 可以访问键控状态和操作符状态，并可以使用定时器来触发事件。ProcessWindowFunction 可以通过实现 ProcessWindowFunction 接口来定义，可以在 WindowedStream 上使用。

在 Flink 中，窗口是将无限流划分为有限大小的块的一种方式。窗口可以基于时间或元素数量进行定义，并可以在流处理过程中动态创建和删除。ProcessWindowFunction 可以在窗口关闭时对窗口中的元素进行聚合操作，并生成输出元素。例如，可以使用 ProcessWindowFunction 来计算窗口中的平均值、最大值、最小值等统计信息。

下面是使用 ProcessWindowFunction 的示例代码。

```java
import org.apache.flink.api.common.functions.MapFunction;
import org.apache.flink.api.common.state.ValueState;
import org.apache.flink.api.common.state.ValueStateDescriptor;
import org.apache.flink.api.java.tuple.Tuple;
import org.apache.flink.api.java.tuple.Tuple2;
import org.apache.flink.streaming.api.datastream.DataStream;
import org.apache.flink.streaming.api.environment.StreamExecutionEnvironment;
import org.apache.flink.streaming.api.functions.KeyedProcessFunction;
import org.apache.flink.util.Collector;

public class StateManagement {
    public static void main(String[] args) throws Exception {
        final StreamExecutionEnvironment env = StreamExecutionEnvironment.getExecutionEnvironment();

        DataStream<Tuple2<String, Integer>> stream = env.socketTextStream("localhost", 9999)
            .map(new MapFunction<String, Tuple2<String, Integer>>() {
                @Override
                public Tuple2<String, Integer> map(String line) throws Exception {
                    String[] tokens = line.split(",");
                    return new Tuple2<>(tokens[0], Integer.parseInt(tokens[1]));
                }
            });

        stream.keyBy(0)
            .process(new KeyedProcessFunction<Tuple, Tuple2<String, Integer>, Tuple2<String, Integer>>() {
                private transient ValueState<Integer> sumState;
```

```java
            @Override
            public void open(org.apache.flink.configuration.Configuration parameters) throws Exception {
                ValueStateDescriptor<Integer> descriptor = new ValueStateDescriptor<>("sum", Integer.class);
                sumState = getRuntimeContext().getState(descriptor);
            }

            @Override
            public void processElement(Tuple2<String, Integer> value, Context ctx, Collector<Tuple2<String, Integer>> out) throws Exception {
                Integer sum = sumState.value();
                if (sum == null) {
                    sum = 0;
                }
                sum += value.f1;
                sumState.update(sum);
                out.collect(new Tuple2<>(value.f0, sum));
            }
        })
        .print();

    env.execute("State Management");
    }
}
```

在这个示例中，我们首先创建了一个 StreamExecutionEnvironment 对象，然后使用 socketTextStream() 方法从本地的 9999 端口读取数据流。接下来，使用 keyBy() 方法将数据流按照第一个元素进行分组，然后使用 process() 方法定义了一个 KeyedProcessFunction，用于对每个分组内的数据进行状态管理。在 process() 方法中，首先使用 ValueStateDescriptor 创建了一个名为"sum"的状态变量，然后在 open() 方法中初始化了这个状态变量。在 processElement() 方法中，首先从状态变量中获取当前的状态值，然后将当前的数据值加入状态值中，并更新状态变量。最后，使用 Collector 将结果输出。

11.6.2 Flink 侧输出流

Flink 侧输出流是指在 Flink 数据流处理过程中，将某些元素输出到一个或多个侧输出流中的机制。侧输出流可以用于将某些元素发送到不同的目的地，例如将异常元素发送到一个异常流中，或将某些元素发送到一个特定的下游处理器中。Flink 中的侧输出流可以通过 OutputTag 来定义，可以在 ProcessFunction、KeyedProcessFunction、CoProcessFunction、KeyedCoProcessFunction、ProcessWindowFunction 和 KeyedProcessWindowFunction 等处理函数中使用。

Flink 侧输出流是一种将数据发送到不同的输出流中的机制，通常用于处理无法处理的数据或错误数据。以下是 Flink 侧输出流并将数据发送到该流中的代码。

```java
import org.apache.flink.streaming.api.datastream.DataStream;
import org.apache.flink.streaming.api.environment.StreamExecutionEnvironment;
import org.apache.flink.streaming.api.functions.ProcessFunction;
import org.apache.flink.util.Collector;
import org.apache.flink.util.OutputTag;

public class FlinkSideOutputExample {
    public static void main(String[] args) throws Exception {
        // 创建执行环境
        StreamExecutionEnvironment env = StreamExecutionEnvironment.getExecutionEnvironment();

        // 从数据源中读取数据
        DataStream<String> input = env.socketTextStream("localhost", 9999);

        // 定义侧输出流标签
        final OutputTag<String> outputTag = new OutputTag<String>("side-output") {};

        // 处理数据并将数据发送到侧输出流
        DataStream<String> result = input.process(new ProcessFunction<String, String>() {
          @Override
           public void processElement(String value, Context ctx, Collector<String> out) throws Exception {
                if (value.startsWith("error:")) {
                    // 将错误数据发送到侧输出流
                    ctx.output(outputTag, value);
                } else {
                    // 将正常数据发送到主输出流
                    out.collect(value);
                }
            }
        });

        // 获取侧输出流
        DataStream<String> sideOutput = result.getSideOutput(outputTag);

        // 输出结果
        result.print();
        sideOutput.print();

        // 执行任务
        env.execute("Flink Side Output Example");
    }
}
```

在这个示例中,我们首先创建了一个执行环境,并从数据源中读取数据。然后,定义了一个侧输出流标签,使用 process 函数处理数据并将数据发送到侧输出流或主输出流。最后,获取侧输出流并输出结果。

11.7 Flink 容错机制

11.7.1 Flink 容错机制概要

Flink 的容错机制是指在 Flink 数据流处理过程中，保证数据处理的正确性和可靠性的机制。Flink 的容错机制主要包括两个方面：检查点和故障恢复。

检查点是指在 Flink 数据流处理过程中，定期将数据流的状态保存到持久化存储中的机制。检查点可以用于在发生故障时恢复数据流的状态，从而保证数据处理的正确性和可靠性。Flink 的检查点机制可以通过 CheckpointConfig 来配置，可以在 DataStream 上使用。

故障恢复是指在 Flink 数据流处理过程中，当发生故障时，自动恢复数据流的状态的机制。Flink 的故障恢复机制可以通过作业管理器和任务管理器之间的协作来实现。当发生故障时，作业管理器会重新启动故障的任务管理器，并从最近的检查点中恢复数据流的状态。

总之，Flink 的容错机制可以保证数据处理的正确性和可靠性，从而保证数据处理的高可用性。

11.7.2 Flink 状态一致性

Flink 状态一致性是指在 Flink 数据流处理过程中，保证状态在不同的任务之间保持一致的机制。Flink 的状态一致性主要包括两个方面：状态后端和状态恢复。

状态后端是指在 Flink 数据流处理过程中，将状态保存到持久化存储中的机制。Flink 支持多种状态后端，包括内存、文件系统、HDFS、RocksDB 等。状态后端可以用于在不同的任务之间共享状态，从而保证状态在不同的任务之间保持一致。

状态恢复是指在 Flink 数据流处理过程中，当发生故障时，自动恢复状态的机制。Flink 的状态恢复机制可以通过检查点来实现。检查点是指在 Flink 数据流处理过程中，定期将数据流的状态保存到持久化存储中的机制。当发生故障时，Flink 会从最近的检查点中恢复状态，从而保证状态在不同的任务之间保持一致。

总之，Flink 的状态一致性可以保证状态在不同的任务之间保持一致，从而保证数据处理的正确性和可靠性。

11.7.3 Flink 容错机制实战

为了实现 Flink 的状态一致性，您需要使用 Flink 的 DataStream API。Flink 的状态一致性是通过检查点机制实现的，它可以在任务失败时恢复任务的状态。以下是 Flink 中实现状态一致性的代码。

```
import org.apache.flink.streaming.api.CheckpointingMode;
import org.apache.flink.streaming.api.datastream.DataStream;
import org.apache.flink.streaming.api.environment.StreamExecutionEnvironment;
```

```java
import org.apache.flink.streaming.api.functions.ProcessFunction;
import org.apache.flink.util.Collector;

public class FlinkStateConsistencyExample {
    public static void main(String[] args) throws Exception {
        // 创建执行环境
        StreamExecutionEnvironment env = StreamExecutionEnvironment.getExecutionEnvironment();

        // 开启检查点机制
        env.enableCheckpointing(5000, CheckpointingMode.EXACTLY_ONCE);

        // 从数据源中读取数据
        DataStream<String> input = env.socketTextStream("localhost", 9999);

        // 处理数据并输出结果
        DataStream<String> result = input.process(new ProcessFunction<String, String>() {
            @Override
            public void processElement(String value, Context ctx, Collector<String> out) throws Exception {
                // 处理数据
                out.collect(value);
            }
        });

        // 输出结果
        result.print();

        // 执行任务
        env.execute("Flink State Consistency Example");
    }
}
```

在这个示例中，我们首先创建了一个执行环境，并开启了检查点机制。然后，从数据源中读取数据，并使用 process 函数处理数据并输出结果。最后，执行任务并启动了 Flink 的状态一致性机制。

11.8 Flink 表和 SQL

11.8.1 Flink 表概述

Flink 表和 SQL 是 Flink 中的一个重要概念，用于对数据流进行处理和分析。Flink 提供了基于表的 API 和 SQL 查询语言，使得用户可以像操作关系型数据库一样对数据流进行查询和分析。

在 Flink 中，表是一种逻辑上的数据结构，可以看作是一张关系型数据库中的表。用户可以通过

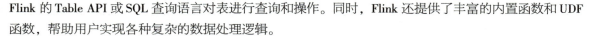

Flink 的 Table API 或 SQL 查询语言对表进行查询和操作。同时，Flink 还提供了丰富的内置函数和 UDF 函数，帮助用户实现各种复杂的数据处理逻辑。

Flink 表和 SQL 的优势在于其简单易用、灵活性强和性能高等特点。首先，Flink 表和 SQL 的语法与关系型数据库类似，用户可以快速上手。其次，Flink 表和 SQL 具有很强的灵活性，用户可以根据自己的需求对数据流进行任意的查询和分析。最后，Flink 表和 SQL 的性能非常高，可以处理大规模的数据流，并且具有很好的容错性和可伸缩性。

11.8.2 Flink 结构化函数

Flink 结构化函数是指在 Flink 数据流处理过程中，对数据进行转换和处理的函数。Flink 结构化函数可以通过 Table API 和 SQL API 来使用，可以在 DataStream 和 DataSet 上使用。

Flink 结构化函数可以将数据流转换为类似于关系型数据库表的结构，包括行和列，可以使用 SQL 语句对数据进行查询和操作。Flink 结构化函数可以支持多种数据源，包括 Kafka、HDFS、文件系统等。Flink 结构化函数可以通过 TableEnvironment 来创建和管理，可以使用 Table API 和 SQL API 来查询和操作数据。Flink 结构化函数具有灵活、高效、易用等特点。

在 Flink 中，常见的结构化函数包括聚合函数、窗口函数、表值函数等。聚合函数可以对数据进行聚合操作，例如求和、平均值、最大值、最小值等。窗口函数可以对数据进行分组和聚合操作，例如滑动窗口、滚动窗口等。表值函数可以将数据流转换为表格形式，例如将 JSON 数据转换为表格形式。

以下是一个使用 Java 在 Flink 中执行结构化操作的示例代码，其中包括定义和使用结构化函数。

11.8.3 Flink 的 SQL 操作

Flink 的 SQL 操作是指在 Flink 数据流处理过程中，对数据进行转换和处理的函数。Flink 结构化操作可以通过 Table API 和 SQL API 来使用，可以在 DataStream 和 DataSet 上使用。Flink 结构化操作可以将数据流转换为类似于关系型数据库表的结构，包括行和列，可以使用 SQL 语句对数据进行查询和操作。Flink 结构化操作可以支持多种数据源，包括 Kafka、HDFS、文件系统等。Flink 结构化操作可以通过 TableEnvironment 来创建和管理，可以使用 Table API 和 SQL API 来查询和操作数据。Flink 结构化操作具有灵活、高效、易用等特点。在 Flink 中，常见的结构化操作包括聚合操作、窗口操作、表值操作等。

以下是使用 Flink Table API 进行聚合操作的示例代码。

```
import org.apache.flink.streaming.api.environment.StreamExecutionEnvironment;
import org.apache.flink.table.api.EnvironmentSettings;
import org.apache.flink.table.api.Table;
import org.apache.flink.table.api.bridge.java.StreamTableEnvironment;

public class FlinkSQL {
    public static void main(String[] args) throws Exception {
```

```java
        final StreamExecutionEnvironment env = StreamExecutionEnvironment.getExecutionEn-
vironment();
        EnvironmentSettings settings = EnvironmentSettings.newInstance().useBlinkPlanner
().inStreamingMode().build();
        StreamTableEnvironment tableEnv = StreamTableEnvironment.create(env, settings);

        String createTable = "CREATE TABLE orders (\n" +
            "   user_id BIGINT,\n" +
            "   product STRING,\n" +
            "   amount INT\n" +
            ") WITH (\n" +
            "   'connector.type' = 'filesystem',\n" +
            "   'connector.path' = 'path/to/orders.csv',\n" +
            "   'format.type' = 'csv'\n" +
            ")";

        String query = "SELECT \n" +
            "   user_id,\n" +
            "   SUM(amount) AS total_amount \n" +
            "FROM orders \n" +
            "GROUP BY user_id";

        tableEnv.executeSql(createTable);
        Table result = tableEnv.sqlQuery(query);
        result.execute().print();
    }
}
```

在这个示例中,我们首先创建了一个 StreamExecutionEnvironment 对象,并使用 EnvironmentSettings 创建了一个 StreamTableEnvironment 对象。然后,使用 CREATE TABLE 语句创建了一个名为"orders"的表,该表包含三个字段:userid、product 和 amount。我们使用' filesystem '连接器将数据从 CSV 文件中读取到表中。接下来,使用 SELECT 语句查询了每个用户的总金额,并使用 GROUP BY 子句按用户 ID 进行分组。最后,使用 execute()方法执行查询并将结果打印出来。

如果您需要模拟数据,可以创建一个名为"orders.csv"的文件,并在其中添加一些订单数据,如下面的模拟数据。

```
1,apple,2
1,banana,3
2,orange,1
2,apple,4
3,banana,2
```

11.9 Flink 复杂事件处理

▶▶ 11.9.1 Flink CEP 简介

1. Flink CEP 概述

Flink CEP（Complex Event Processing）是指在 Flink 数据流处理过程中，对复杂事件进行处理的一种技术。Flink CEP 可以通过定义模式来识别复杂事件，模式可以包括多个事件，可以定义事件之间的关系和时间限制。

Flink CEP 可以支持多种数据源，包括 Kafka、HDFS、文件系统等。

Flink CEP 可以通过 DataStream API 和 Table API 来使用，可以对数据流进行实时处理和分析。

Flink CEP 可以应用于多种场景，例如金融交易、物联网、网络安全等。

2. Flink CEP 原理

Flink CEP 模式的原理是基于有限状态自动机（Finite State Machine，FSM）的。在 Flink CEP 中，每个模式都可以表示为一个有限状态自动机。这个自动机由一组状态和转移函数组成，其中状态表示事件序列的不同阶段，转移函数表示事件序列从一个状态转移到另一个状态的条件。

3. Flink CEP 处理流程

Flink CEP 模式的处理过程可以分为以下几个步骤。

1）定义模式：定义一个符合要求的事件序列模式，可以使用 Flink CEP 提供的 API 进行定义。

2）构建自动机：将定义好的模式转换为有限状态自动机。

3）匹配事件：将流数据中的事件与自动机进行匹配，如果匹配成功，则触发相应的处理逻辑。

4）输出结果：将处理后的结果输出到指定的目的地。

4. Flink CEP 应用场景

Flink CEP 的应用场景非常广泛，以下是一些常见的应用场景。

1）金融交易监控：Flink CEP 可以用于监控金融交易数据流，以检测潜在的欺诈行为或异常交易。例如，可以定义一个模式，表示在短时间内发生了多次高额交易，这可能是一个欺诈行为的指标。当 Flink CEP 检测到这个模式时，可以触发相应的警报或采取其他的行动。

2）物联网数据分析：Flink CEP 可以用于处理物联网设备生成的实时数据流，以检测和识别特定的事件模式。例如，可以定义一个模式，表示在某个时间段内，某个设备的温度超过了一定的阈值。当 Flink CEP 检测到这个模式时，可以触发相应的行动，例如发送警报或采取措施来调整设备的设置。

3）电信网络监控：Flink CEP 可以用于监控电信网络数据流，以检测和识别特定的事件模式。例如，可以定义一个模式，表示在某个时间段内，某个区域的网络流量超过了一定的阈值。当 Flink CEP 检测到这个模式时，可以触发相应的行动，例如调整网络带宽或采取其他的措施来优化网络性能。

总之，Flink CEP 可以用于处理各种实时数据流，并在发现特定的事件模式时采取相应的行动。

它可以帮助企业实时监控业务流程，提高业务效率和安全性。

11.9.2 Flink CEP 个体模式

Flink CEP 的个体模式是指对单个事件进行匹配和处理，可以通过定义规则来识别事件，规则可以包括多个条件和时间限制。Flink CEP 的个体模式可以应用于多种场景，例如异常检测、实时监控等。

以下是生成 Flink CEP 模拟数据的代码。

```java
import org.apache.flink.streaming.api.watermark.Watermark;
import java.util.Random;

public class ClickEventSource implements SourceFunction<ClickEvent> {
    private volatile boolean running = true;

    @Override
    public void run(SourceContext<ClickEvent> ctx) throws Exception {
        Random rand = new Random();
        long timestamp = System.currentTimeMillis();
        while (running) {
            ClickEvent event = new ClickEvent(
                rand.nextInt(10), // user ID
                rand.nextInt(5), // page ID
                rand.nextInt(3), // button ID
                timestamp
            );
            ctx.collectWithTimestamp(event, timestamp);
            ctx.emitWatermark(new Watermark(timestamp));
            timestamp += rand.nextInt(100);
            Thread.sleep(10);
        }
    }

    @Override
    public void cancel() {
        running = false;
    }
}
```

以下是使用 Flink CEP 的程序示例，用于检测连续三次单击 ID 为 2 的按钮的模式代码。

```java
import org.apache.flink.api.common.functions.MapFunction;
import org.apache.flink.cep.CEP;
import org.apache.flink.cep.PatternStream;
import org.apache.flink.cep.pattern.Pattern;
import org.apache.flink.cep.pattern.conditions.IterativeCondition;
import org.apache.flink.streaming.api.datastream.DataStream;
```

```java
import org.apache.flink.streaming.api.environment.StreamExecutionEnvironment;
import org.apache.flink.streaming.api.windowing.time.Time;
import java.util.List;
import java.util.Map;

public class ClickEventCEP {

    public static void main(String[] args) throws Exception {
        StreamExecutionEnvironment env = StreamExecutionEnvironment.getExecutionEnvironment();

        DataStream<ClickEvent> input = env.addSource(new ClickEventSource());

        Pattern<ClickEvent, ? > pattern = Pattern.<ClickEvent>begin("start")
            .where(new IterativeCondition<ClickEvent>() {
                @Override
                public boolean filter(ClickEvent event, Context<ClickEvent> context) throws Exception {
                    return event.getButtonId() == 2;
                }
            })
            .followedBy("middle")
            .where(new IterativeCondition<ClickEvent>() {
                @Override
                public boolean filter(ClickEvent event, Context<ClickEvent> context) throws Exception {
                    return event.getButtonId() == 2;
                }
            })
            .followedBy("end")
            .where(new IterativeCondition<ClickEvent>() {
                @Override
                public boolean filter(ClickEvent event, Context<ClickEvent> context) throws Exception {
                    return event.getButtonId() == 2;
                }
            })
            .within(Time.seconds(10));

        PatternStream<ClickEvent> patternStream = CEP.pattern(input, pattern);

        DataStream<String> result = patternStream.select(
            (Map<String, List<ClickEvent>> patternMatch) -> {
                List<ClickEvent> startEvents = patternMatch.get("start");
                List<ClickEvent> middleEvents = patternMatch.get("middle");
                List<ClickEvent> endEvents = patternMatch.get("end");
                return "Found pattern: " + startEvents.size() + " " + middleEvents.size() + " " + endEvents.size();
```

```
            }
        );

        result.print();
        env.execute("ClickEventCEP");
    }
}
```

在此示例中,我们定义了一个模式,该模式在 10 秒窗口内匹配 ID 为 2 的按钮连续三次点击。然后,使用 CEP.pattern( )方法将此模式应用于 ClickEvent 对象的输入流。最后,选择匹配的模式并将其打印到控制台。

要为此程序生成模拟数据,您可以使用前面示例中提供的 ClickEventSource 类。只需调用 env.addSource(new ClickEventSource( ))即可生成具有随机用户 ID、页面 ID 和按钮 ID 的 ClickEvent 对象流。

以下是生成模拟数据的示例代码。

```
StreamExecutionEnvironment env = StreamExecutionEnvironment.getExecutionEnvironment();
DataStream<ClickEvent> input = env.addSource(new ClickEventSource());
input.print();
env.execute("ClickEventSource");
```

在此示例中,我们创建了一个 ClickEventSource 对象,并将其添加到 Flink 的执行环境中。然后,打印生成的 ClickEvent 对象流并执行程序。

▶▶ 11.9.3　Flink CEP 组合模式

Flink CEP(Complex Event Processing)是一种用于处理复杂事件的框架。组合模式是 Flink CEP 中的一种模式,它由多个简单模式组成,这些简单模式可以按照特定的顺序和时间间隔组合在一起。在 Flink CEP 中,组合模式可以使用 Pattern 类定义,其中每个简单模式可以使用 IterativeCondition 类定义。组合模式可以应用于数据流中,使用 CEP.pattern 方法进行匹配,并使用 PatternSelectFunction 处理匹配到的事件序列。

为了使用 Flink CEP 组合模式,可以使用前面提供的示例代码作为基础,并对其进行修改,以匹配我们的需求。以下是一个示例代码,用于检测连续两次点击 ID 为 1 的按钮的模式。

```
import org.apache.flink.api.common.functions.MapFunction;
import org.apache.flink.cep.CEP;
import org.apache.flink.cep.PatternStream;
import org.apache.flink.cep.pattern.Pattern;
import org.apache.flink.cep.pattern.conditions.IterativeCondition;
import org.apache.flink.streaming.api.datastream.DataStream;
import org.apache.flink.streaming.api.environment.StreamExecutionEnvironment;
import org.apache.flink.streaming.api.windowing.time.Time;

import java.util.List;
import java.util.Map;
```

```java
public class ClickEventCEP {

    public static void main(String[] args) throws Exception {
        StreamExecutionEnvironment env = StreamExecutionEnvironment.getExecutionEnvironment();

        DataStream<ClickEvent> input = env.addSource(new ClickEventSource());

        Pattern<ClickEvent, ? > pattern = Pattern.<ClickEvent>begin("start")
                .where(new IterativeCondition<ClickEvent>() {
                    @Override
                    public boolean filter(ClickEvent event, Context<ClickEvent> context)
throws Exception {
                        return event.getButtonId() == 1;
                    }
                })
                .followedBy("end")
                .where(new IterativeCondition<ClickEvent>() {
                    @Override
                    public boolean filter(ClickEvent event, Context<ClickEvent> context)
throws Exception {
                        return event.getButtonId() == 1;
                    }
                })
                .within(Time.seconds(10));

        PatternStream<ClickEvent> patternStream = CEP.pattern(input, pattern);

        DataStream<String> result = patternStream.select(
                (Map<String, List<ClickEvent>> patternMatch) -> {
                    List<ClickEvent> startEvents = patternMatch.get("start");
                    List<ClickEvent> endEvents = patternMatch.get("end");
                    return "Found pattern: " + startEvents.size() + " " + endEvents.size();
                }
        );

        result.print();

        env.execute("ClickEventCEP");
    }
}
```

在此示例中，我们定义了一个模式，该模式在 10 秒窗口内匹配 ID 为 1 的按钮连续两次点击。然后，使用 CEP.pattern( ) 方法将此模式应用于 ClickEvent 对象的输入流。最后，选择匹配的模式并将其打印到控制台。

要为此程序生成模拟数据，您可以使用前面示例中提供的 ClickEventSource 类。只需调用

env.addSource（new ClickEventSource()）即可生成具有随机用户 ID、页面 ID 和按钮 ID 的 ClickEvent 对象流。

以下是生成模拟数据的示例代码。

```
StreamExecutionEnvironment env = StreamExecutionEnvironment.getExecutionEnvironment();
DataStream<ClickEvent> input = env.addSource(new ClickEventSource());
input.print();
env.execute("ClickEventSource");
```

在此示例中，我们创建了一个 ClickEventSource 对象并将其添加到 Flink 的执行环境中。然后，打印生成的 ClickEvent 对象流并执行程序。

11.9.4 Flink CEP 超时事件提取

Flink CEP 中的超时事件提取是一种将数据流中的事件序列与预定义的模式进行匹配的机制。在 Flink CEP 中，可以使用 within 方法定义一个时间范围，如果在该时间范围内没有匹配到模式的下一个事件，则会触发超时事件。在 Java 中，可以使用 Flink CEP 库来实现超时事件提取。

为了使用 Flink CEP 超时事件，可以使用前面提供的示例代码作为基础，并对其进行修改，以匹配我们的需求。以下是一个示例代码，用于检测在 5 秒内没有任何点击事件的模式。

```java
import org.apache.flink.api.common.functions.MapFunction;
import org.apache.flink.cep.CEP;
import org.apache.flink.cep.PatternStream;
import org.apache.flink.cep.pattern.Pattern;
import org.apache.flink.streaming.api.datastream.DataStream;
import org.apache.flink.streaming.api.environment.StreamExecutionEnvironment;
import org.apache.flink.streaming.api.windowing.time.Time;
import java.util.List;
import java.util.Map;

public class ClickEventCEP {

    public static void main(String[] args) throws Exception {
        StreamExecutionEnvironment env = StreamExecutionEnvironment.getExecutionEnvironment();

        DataStream<ClickEvent> input = env.addSource(new ClickEventSource());

        Pattern<ClickEvent, ? > pattern = Pattern.<ClickEvent>begin("start")
            .where(new IterativeCondition<ClickEvent>() {
                @Override
                public boolean filter(ClickEvent event, Context<ClickEvent> context) throws Exception {
                    return true;
                }
            })
            .notFollowedBy("end")
```

```
            .where(new IterativeCondition<ClickEvent>() {
                @Override
                public boolean filter(ClickEvent event, Context<ClickEvent> context) throws Exception {
                    return true;
                }
            })
            .within(Time.seconds(5));

        PatternStream<ClickEvent> patternStream = CEP.pattern(input, pattern);

        DataStream<String> result = patternStream.select(
            (Map<String, List<ClickEvent>> patternMatch) -> {
                List<ClickEvent> startEvents = patternMatch.get("start");
                return "Found pattern: " + startEvents.size();
            }
        );

        result.print();
        env.execute("ClickEventCEP");
    }
}
```

在此示例中,我们定义了一个模式,该模式在 5 秒窗口内匹配没有任何点击事件。然后,使用 **CEP.pattern( )** 方法将此模式应用于 **ClickEvent** 对象的输入流。最后,选择匹配的模式并将其打印到控制台。

要为此程序生成模拟数据,您可以使用前面示例中提供的 **ClickEventSource** 类。只需调用 **env.addSource(new ClickEventSource( ))** 即可生成具有随机用户 ID、页面 ID 和按钮 ID 的 **ClickEvent** 对象流。

以下是生成模拟数据的示例代码。

```
StreamExecutionEnvironment env = StreamExecutionEnvironment.getExecutionEnvironment();
DataStream<ClickEvent> input = env.addSource(new ClickEventSource());
input.print();
env.execute("ClickEventSource");
```

在此示例中,我们创建了一个 ClickEventSource 对象并将其添加到 Flink 的执行环境中。然后,打印生成的 ClickEvent 对象流并执行程序。

第12章

电商推荐系统实战

本章内容介绍推荐系统的基本概念和工作原理,包括协同过滤、内容过滤、混合推荐等。学习推荐系统的数据处理和特征工程,包括数据清洗、特征提取、特征选择等。掌握推荐系统的常用算法和模型,包括基于邻域的算法。学习推荐系统的评估和优化方法,包括离线评估、在线评估、实践推荐系统的应用案例。

在这个实战项目中,我们将使用 Spark 构建一个高效的电商推荐系统。Spark 是一个大规模数据处理工具,具有强大的分布式计算能力,能够处理海量的数据集。我们将利用其提供的功能和算法,结合电商推荐系统的实际需求,实现一个实用的推荐系统。

12.1 推荐系统的概述

推荐系统是通过挖掘用户与项目之间的关系,帮助用户从大量数据中发现其可能感兴趣的项目,如网页、服务、商品、视频等,并生成个性化推荐,以满足个性化需求。

本节介绍推荐系统在不同行业和领域的应用场景,并探讨如何设计和构建一个高效的推荐系统。通过本实战,你将了解推荐系统的核心概念、算法原理和应用领域,并掌握如何将推荐系统应用于实际业务问题。推荐系统在许多场景中都有广泛的应用。以下是一些典型的应用场景。

1)电子商务网站:这是推荐系统最常用的场景之一。例如,当你在购物网站上搜索某一款产品时,推荐系统可以根据你的搜索历史和购买行为,为你推荐一些相关的产品。

2)音乐和视频流媒体平台:这些平台通常使用推荐系统来推荐新的音乐或视频给用户。例如,根据你的听歌历史和浏览记录,推荐系统可以为你推荐类似的音乐或视频。

3)社交网络:社交网络使用推荐系统来推荐新的朋友或群组。例如,根据你的朋友网络和兴趣爱好,推荐系统可以为你推荐一些可能想关注的人或群组。

4)新闻和阅读平台:这些平台使用推荐系统来推荐新的文章或故事。例如,根据你的阅读历史和兴趣爱好,推荐系统可以为你推荐一些可能想阅读的新闻或文章。

5)广告系统:广告平台可以使用推荐系统来预测你对哪些广告可能感兴趣,并向你推荐相关的广告。例如,根据你在电子商务网站上的购买历史,推荐系统可以为你推荐一些相关的产品广告。

总之，推荐系统在许多领域都有广泛的应用，可以帮助用户更好地发现他们可能感兴趣的内容或产品，也可以帮助平台提高用户满意度和增加收益。

12.1.1 推荐系统算法

推荐系统的核心算法包括协同过滤、基于内容的推荐、混合推荐等。协同过滤是基于用户或物品的相似性进行推荐的算法；基于内容的推荐是根据物品本身的属性进行推荐的算法；混合推荐则是将前两种算法结合起来，形成更加准确的推荐结果。

1. 基于用户 CF 和基于物品 CF

基于用户的协同过滤（User CF）是一种常见的推荐算法，其主要步骤如下。

1）基于用户对物品的偏好找到相邻用户。

2）将相邻用户喜欢的推荐给当前用户。

3）找到 K 个邻居后，根据邻居的相似度权重以及他们对物品的偏好，预测当前用户没有偏好的未涉及物品，计算得到一个排序的物品列表作为推荐。

计算上，就是将一个用户对所有物品的偏好作为一个向量来计算用户之间的相似度。

对于用户 A，根据用户的历史偏好，这里只计算得到一个相邻用户 C，然后将用户 C 喜欢的物品 D 推荐给用户 A，如表 12-1 所示。

表 12-1　基于用户推荐的示意图

序号	用户/物品	物品 A	物品 B	物品 C	物品 D
1	用户 A	√	—	√	推荐
2	用户 B	—	√	—	—
3	用户 C	√	—	√	√

2. 基于物品的协同过滤（Item CF）

基于物品的协同过滤（Item CF）是一种常见的推荐算法，其主要步骤如下。

1）基于用户对物品的偏好找到相似的物品。

2）根据用户的历史偏好，推荐相似的物品给他。

3）得到物品的相似物品后，根据用户的历史偏好预测当前用户还没有表示偏好的物品，计算得到一个排序的物品列表作为推荐。

从计算的角度看，就是将所有用户对某个物品的偏好作为一个向量来计算物品之间的相似度。

对于物品 A，根据所有用户的历史偏好，喜欢物品 A 的用户都喜欢物品 C，得出物品 A 和物品 C 比较相似，而用户 C 喜欢物品 A，那么可以推断出用户 C 可能也喜欢物品 C，如表 12-2 所示。

表 12-2 基于物品推荐的示意图

序 号	用户/物品	物品 A	物品 B	物品 C
1	用户 A	√	—	√
2	用户 B	√	√	√
3	用户 C	√	—	推荐

3. 基于 Spark MLLib 的协同过滤

（1）基于用户（User-Based）的协同过滤

在 Spark MLLib 中，通过读取（用户-物品-评分）数据可以构建相似性矩阵。利用相似性矩阵计算一些用户的指标，包括用户对每种物品的评分、用户对所有物品的平均评分、用户对一种商品的评分、某个用户相对于其他用户的相似性。

（2）基于物品（Item-Based）的协同过滤

基本的逻辑是首先得到某个用户评价过（买过）的商品，然后计算其他商品与该商品的相似度并排序，从高到低把不在用户评价过商品里的其他商品推荐给用户。

12.1.2 推荐系统的数据

偏好值就是用户对物品的喜爱程度，推荐系统所做的事就是根据这些数据为用户推荐还没有见过的物品，并且猜测这个物品用户喜欢的概率比较大。

用户 ID 和物品 ID 一般通过系统的业务数据库就可以获得，偏好值的采集一般会有很多办法，比如评分、投票、转发、保存书签、页面停留时间等，然后系统根据用户的这些行为流水，采取减噪、归一化、加权等方法综合给出偏好值。一般不同的业务系统给出偏好值的计算方法不一样。常见的用户行为和偏好值统计如表 12-3 所示。

表 12-3 用户行为和偏好值

序号	用户行为	类型	特 征	作 用
1	评分	显式	整数量化的偏好，可能的取值是 [0, n]；n 一般取值为 5 或者是 10	通过用户对物品的评分，可以精确地得到用户的偏好
2	投票	显式	布尔量化的偏好，取值是 0 或 1	通过用户对物品的投票，可以较精确地得到用户的偏好
3	转发	显式	布尔量化的偏好，取值是 0 或 1	通过用户对物品的投票，可以精确地得到用户的偏好；如果是站内，同时可以推理得到被转发人的偏好
4	保存书签	显式	布尔量化的偏好，取值是 0 或 1	通过用户对物品的投票，可以精确地得到用户的偏好
5	标记标签	显式	对于一些单词，需要对单词进行分析，得到偏好	通过分析用户的标签，可以得到用户对项目的理解，同时可以分析出用户的情感：喜欢还是讨厌
6	评论	显式	对于一段文字，需要进行文本分析，得到偏好	通过分析用户的标签，可以得到用户对项目的理解，同时可以分析出用户的情感：喜欢还是讨厌

（续）

序号	用户行为	类型	特 征	作 用
7	点击查看	隐式	一组用户的点击，用户对物品感兴趣，需要进行分析，得到偏好	用户的点击一定程度上反映了用户的注意力，所以它也可以从一定程度上反映用户的喜好
8	页面停留时间	隐式	一组时间信息，噪声大，需要进行去噪、分析，得到偏好	用户的页面停留时间一定程度上反映了用户的注意力和喜好，但噪声偏大，不好利用
9	购买	隐式	布尔量化的偏好，取值是 0 或 1	用户的购买很明确地说明对这个项目感兴趣

12.2 电商推荐系统架构

12.2.1 电商推荐系统模块

1. 用户可视化

主要负责实现和用户的交互以及业务数据的展示，主要是前端开发实现，部署在服务上。

2. 业务服务

主要实现 JavaEE 层面整体的业务逻辑，通过 Spring 进行构建，对接业务需求，后端开发（Java 开发）实现，部署在 Tomcat 上。

3. 数据存储部分

业务数据库：项目采用广泛应用的文档数据库 MongoDB 作为主数据库，主要负责平台业务逻辑数据的存储。

缓存数据库：项目采用 Redis 作为缓存数据库，主要用来支撑实时推荐系统部分对于数据的高速获取需求。

4. 离线推荐部分

离线统计服务：批处理统计性业务采用 Spark Core + Spark SQL 进行实现，实现对指标类数据的统计任务。

离线推荐服务：离线推荐业务采用 Spark Core + Spark MLlib 进行实现，采用 ALS 算法进行实现。

5. 实时推荐部分

日志采集服务：通过利用 Flume-ng 对业务平台中用户对于商品的一次评分行为进行采集，实时发送到 Kafka 集群。

消息缓冲服务：项目采用 Kafka 作为流式数据的缓存组件，接受来自 Flume 的数据采集请求，并将数据推送到项目的实时推荐系统部分。

实时推荐服务：项目采用 Spark Streaming 作为实时推荐系统，通过接收 Kafka 中缓存的数据，通过设计的推荐算法实现对实时推荐的数据处理，并将结构合并更新到 MongoDB 数据库。

12.2.2 创建一个推荐项目

1. 项目介绍

新建一个推荐系统项目，并且将系统名称命名为 jareny-recommen，推荐系统分为导入数据模块、离线统计模块、离线推荐模块、在线推荐模块、基于内容推荐。下面详细介绍每一个模块的实现。

2. 导入公共依赖

在推荐系统项目的 pom.xml 文件中添加依赖，如下。

```xml
<?xml version="1.0" encoding="UTF-8"?>
<project xmlns="http://maven.apache.org/POM/4.0.0"
     xmlns:xsi="http://www.w3.org/2001/XMLSchema-instance"
     xsi:schemaLocation="http://maven.apache.org/POM/4.0.0 http://maven.apache.org/xsd/maven-4.0.0.xsd">
    <modelVersion>4.0.0</modelVersion>

    <groupId>com.it.jareny.bigdata</groupId>
    <artifactId>jareny-recommend</artifactId>
    <packaging>pom</packaging>
    <version>1.0-SNAPSHOT</version>
    <modules>
        <module>dataLoader</module>
        <module>statisticsRecommender</module>
        <module>offlineRecommender</module>
        <module>onlineRecommender</module>
        <module>model</module>
        <module>common</module>
    </modules>

    <properties>
        <scala.version>2.11</scala.version>
        <spark.version>2.4.5</spark.version>
    </properties>

    <dependencies>
        <!--Spark 依赖-->
        <!-- https://mvnrepository.com/artifact/org.apache.spark/spark-core_2.11 -->
        <dependency>
            <groupId>org.apache.spark</groupId>
            <artifactId>spark-core_${scala.version}</artifactId>
            <version>${spark.version}</version>
        </dependency>

        <!-- https://mvnrepository.com/artifact/org.apache.spark/spark-sql_2.11 -->
        <dependency>
```

```xml
        <groupId>org.apache.spark</groupId>
        <artifactId>spark-sql_${scala.version}</artifactId>
        <version>${spark.version}</version>
</dependency>

<dependency>
        <groupId>org.apache.spark</groupId>
        <artifactId>spark-streaming_${scala.version}</artifactId>
        <version>${spark.version}</version>
</dependency>

<dependency>
        <groupId>org.apache.spark</groupId>
        <artifactId>spark-mllib_${scala.version}</artifactId>
        <version>${spark.version}</version>
</dependency>

<dependency>
        <groupId>org.apache.spark</groupId>
        <artifactId>spark-streaming-kafka-0-10_${scala.version}</artifactId>
        <version>${spark.version}</version>
</dependency>

<dependency>
        <groupId>org.apache.spark</groupId>
        <artifactId>spark-streaming-kafka-0-10-assembly_${scala.version}</artifactId>
        <version>${spark.version}</version>
</dependency>

<dependency>
        <groupId>org.jblas</groupId>
        <artifactId>jblas</artifactId>
        <version>1.2.4</version>
</dependency>

<!-- MongoDB 依赖 -->
<dependency>
        <groupId>org.mongodb.spark</groupId>
        <artifactId>mongo-spark-connector_2.11</artifactId>
        <version>2.4.1</version>
</dependency>
<dependency>
        <groupId>org.mongodb</groupId>
        <artifactId>mongo-java-driver</artifactId>
        <version>3.12.7</version>
</dependency>
```

```xml
        <dependency>
            <groupId>cn.hutool</groupId>
            <artifactId>hutool-all</artifactId>
            <version>5.6.3</version>
            <exclusions>
                <exclusion>
                    <artifactId>mongo-java-driver</artifactId>
                    <groupId>org.mongodb</groupId>
                </exclusion>
            </exclusions>
        </dependency>

        <dependency>
            <groupId>org.projectlombok</groupId>
            <artifactId>lombok</artifactId>
            <version>1.18.24</version>
        </dependency>
    </dependencies>

</project>
```

3. 创建公共模型模块

新建 model 模块，在 model 模块的 pom.xml 文件中添加依赖，如下。

```xml
<?xml version="1.0" encoding="UTF-8"?>
<project xmlns="http://maven.apache.org/POM/4.0.0"
         xmlns:xsi="http://www.w3.org/2001/XMLSchema-instance"
         xsi:schemaLocation="http://maven.apache.org/POM/4.0.0 http://maven.apache.org/xsd/maven-4.0.0.xsd">
    <parent>
        <artifactId>jareny-recommend</artifactId>
        <groupId>com.it.jareny.bigdata</groupId>
        <version>1.0-SNAPSHOT</version>
    </parent>
    <modelVersion>4.0.0</modelVersion>

    <artifactId>model</artifactId>

</project>
```

1）创建商品对象，用于解析数据集，并且存储到数据库。

```java
package com.jareny.bigdata.model;

import java.io.Serializable;

public class Product implements Serializable {
```

```java
// productId   商品 id
private int productId;
// name        商品名
private String name;
// mageUrl     商品图片 URL
private String imageUrl;
//categories   商品分类
private String categories;
//tags         用户 UGC 标签
private String tags;

public Product() {
}

public Product(int productId, String name, String imageUrl, String categories, String tags) {
    this.productId = productId;
    this.name = name;
    this.imageUrl = imageUrl;
    this.categories = categories;
    this.tags = tags;
}

public int getProductId() {
    return productId;
}

public void setProductId(int productId) {
    this.productId = productId;
}

public String getName() {
    return name;
}

public void setName(String name) {
    this.name = name;
}

public String getImageUrl() {
    return imageUrl;
}

public void setImageUrl(String imageUrl) {
    this.imageUrl = imageUrl;
}
```

```java
    public String getCategories() {
        return categories;
    }

    public void setCategories(String categories) {
        this.categories = categories;
    }

    public String getTags() {
        return tags;
    }

    public void setTags(String tags) {
        this.tags = tags;
    }
}
```

2)创建用户评分对象,用于解析数据集,并且存储到数据库。

```java
package com.jareny.bigdata.model;

import java.io.Serializable;

public class UserRating implements Serializable {
    // 用户 id
    private int userId;
    // 商品 id
    private int productId;
    // 用户评分
    private double score;
    // 用户评分时间
    private int timestamp;

    public UserRating() {
    }

    public UserRating(int userId, int productId, double score, int timestamp) {
        this.userId = userId;
        this.productId = productId;
        this.score = score;
        this.timestamp = timestamp;
    }

    public int getUserId() {
        return userId;
    }
```

```java
    public void setUserId(int userId) {
        this.userId = userId;
    }

    public int getProductId() {
        return productId;
    }

    public void setProductId(int productId) {
        this.productId = productId;
    }

    public double getScore() {
        return score;
    }

    public void setScore(double score) {
        this.score = score;
    }

    public int getTimestamp() {
        return timestamp;
    }

    public void setTimestamp(int timestamp) {
        this.timestamp = timestamp;
    }
}
```

3) 定义推荐对象,用于标准推荐列表数据。

```java
package com.jareny.bigdata.model;

import java.io.Serializable;

// 定义标准推荐对象
public class Recommendation implements Serializable {
    private int productId;
    private double score;

    public Recommendation() {
    }

    public Recommendation(int productId, double score) {
        this.productId = productId;
        this.score = score;
    }
```

```java
    public int getProductId() {
        return productId;
    }

    public void setProductId(int productId) {
        this.productId = productId;
    }

    public double getScore() {
        return score;
    }

    public void setScore(double score) {
        this.score = score;
    }
}
```

4)定义用户相似推荐对象,用于为用户推荐商品。

```java
package com.jareny.bigdata.model;

import java.io.Serializable;
import java.util.List;

//定义用户的推荐列表
public class UserRecs implements Serializable {

    private int userId;
    private List<Recommendation> recs;

    public UserRecs() {
    }

    public UserRecs(int userId, List<Recommendation> recs) {
        this.userId = userId;
        this.recs = recs;
    }

    public int getUserId() {
        return userId;
    }

    public void setUserId(int userId) {
        this.userId = userId;
    }
```

```java
    public List<Recommendation> getRecs() {
        return recs;
    }

    public void setRecs(List<Recommendation> recs) {
        this.recs = recs;
    }
}
```

5）定义商品相似对象，用于商品相似度推荐。

```java
package com.jareny.bigdata.model;

import java.io.Serializable;
import java.util.List;

public class ProductRecs implements Serializable {

    private int productId;

    private List<Recommendation> recs;

    public ProductRecs() {
    }

    public ProductRecs(int productId, List<Recommendation> recs) {
        this.productId = productId;
        this.recs = recs;
    }

    public int getProductId() {
        return productId;
    }

    public void setProductId(int productId) {
        this.productId = productId;
    }

    public List<Recommendation> getRecs() {
        return recs;
    }

    public void setRecs(List<Recommendation> recs) {
        this.recs = recs;
    }
}
```

4. 创建公共工具模块

新建 common 模块，在 common 模块的 pom.xml 文件中添加依赖，如下。

```xml
<?xml version="1.0" encoding="UTF-8"? >
<project xmlns="http://maven.apache.org/POM/4.0.0"
     xmlns:xsi="http://www.w3.org/2001/XMLSchema-instance"
      xsi:schemaLocation="http://maven.apache.org/POM/4.0.0 http://maven.apache.org/xsd/maven-4.0.0.xsd">
    <parent>
        <artifactId>jareny-recommend</artifactId>
        <groupId>com.it.jareny.bigdata</groupId>
        <version>1.0-SNAPSHOT</version>
    </parent>
    <modelVersion>4.0.0</modelVersion>

    <artifactId>common</artifactId>

</project>
```

1) 创建 MongoDB 数据访问工具类。

```java
package com.jareny.bigdata.util;

import com.google.common.collect.Maps;
import com.mongodb.spark.MongoSpark;
import com.mongodb.spark.config.ReadConfig;
import com.mongodb.spark.config.WriteConfig;
import org.apache.spark.api.java.JavaRDD;
import org.apache.spark.api.java.JavaSparkContext;
import org.apache.spark.sql.Dataset;
import org.apache.spark.sql.Row;
import org.bson.Document;

import java.util.HashMap;
import java.util.Map;

public class MongoUtil {

    /**
     * Spark 从 MongoDB 数据库读取数据
     *
     * @param javaSparkContext Spark 配置中需要设定 mongodb 的数据库连接
     * @return
     */
    public Dataset<Row> load(JavaSparkContext javaSparkContext) {
        return MongoSpark.load(javaSparkContext).toDF();
```

```java
    }

    /**
     * 保存数据到 MongoDB 数据库
     *
     * @param javaSparkContext Spark 配置
     * @param config           mongodb 的数据库连接
     * @return
     */
    public Dataset<Row> load(JavaSparkContext javaSparkContext, Map<String, String>
config) {
        return MongoSpark.load(javaSparkContext, ReadConfig.create(javaSparkContext).
withOptions(config)).toDF();
    }

    /**
     * Spark 保存到 mongodb 数据库
     *
     * @param sparkContext    spark 配置
     * @param documentRdd     存储的到 mongodb 的 Rdd
     * @param collectionName  存储集合名称
     */
    public static void saveToMongo(JavaSparkContext sparkContext,
                                   JavaRDD<Document> documentRdd,
                                   String collectionName) {

        // 1. 配置 MongoDB 写入参数
        HashMap<String, String> writeOverrides = Maps.newHashMap();
        writeOverrides.put("collection", collectionName);
        writeOverrides.put("writeConcern.w", "majority");
        WriteConfig writeConfig = WriteConfig.create(sparkContext).withOptions(writeOver-
rides);

        // 2. 保存到数据库
        MongoSpark.save(documentRdd, writeConfig);
    }
}
```

2）创建数据转换工具类。

```
package com.jareny.bigdata.util;

import cn.hutool.core.bean.BeanUtil;
import cn.hutool.json.JSONUtil;
import org.apache.spark.api.java.JavaRDD;
import org.apache.spark.api.java.function.Function;
import org.apache.spark.sql.Dataset;
```

```java
import org.apache.spark.sql.Encoders;
import org.apache.spark.sql.Row;
import org.bson.Document;

/**
 * 转换函数
 */
public class ConvertUtil {

    /**
     * avaRDD<String> 转换 JavaRDD<Document> 函数
     *
     * @param stringRdd JavaRDD 是 String 类型的
     * @param function  转换函数
     * @param <T>       实体类的类型
     * @return
     */
    public static <T> JavaRDD<Document> convertToDocument(JavaRDD<String> stringRdd, Function<String, T> function) {
        // 1.String 类型 RDD 调用 map 函数转换成实体类的 RDD
        JavaRDD<T> entityRdd = stringRdd.map(function);
        // 2.实体类的 RDD 调用 map 函数转换成 MongoDB 中 Document 的 RDD
        return entityRdd.map(document -> Document.parse(JSONUtil.toJsonStr(document)));
    }

    /**
     * Dataset<Row> 转成 JavaRDD<Document>
     *
     * @param dataset
     * @param clazz
     * @param <T>
     * @return
     */
    public static <T> JavaRDD<Document> convertToDocument(Dataset<Row> dataset, Class<T> clazz) {
        // 转成实体类
        JavaRDD<T> javaRDD = convertToModel(dataset, clazz);
        // 转成 Document
        return javaRDD.map(document -> Document.parse(JSONUtil.toJsonStr(document)));
    }

    /**
     * Dataset<Row> 转成 JavaRDD<T>
     *
     * @param dataset
     * @param clazz
```

```
    * @param <T>
    * @return
    */
   public static <T> JavaRDD<T> convertToModel(Dataset<Row> dataset, Class<T> clazz) {
       JavaRDD<T> javaRDD = dataset.as(Encoders.bean(clazz.getClass()))
               .javaRDD()
               .map(rating -> BeanUtil.copyProperties(rating, clazz));
       return javaRDD;
   }

   /**
    * Dataset<Row> 转成 JavaRDD<T>
    *
    * @param dataset
    * @param clazz
    * @param <T>
    * @return
    */
   public static <U,T> JavaRDD<T> convertToModel(Dataset<Row> dataset, Class<U> clazz,
Function<U,T> function) {
       JavaRDD<T> javaRDD = dataset.as(Encoders.bean(clazz))
               .javaRDD()
               .map(function::call);
       return javaRDD;
   }
}
```

12.3 加载数据模块

12.3.1 模块介绍

加载数据模块就是将采集到 HDFS 的数据，使用 Spark 处理成有数据格式的数据集。然后将数据写到数据库中，便于系统做数据统计和分析。

12.3.2 模块实现

1. 新建加载数据模块并且添加依赖

新建 dataLoader 模块，在 dataLoader 模块的 pom.xml 文件中添加依赖，如下。

```
<?xml version="1.0" encoding="UTF-8"?>
<project xmlns="http://maven.apache.org/POM/4.0.0"
    xmlns:xsi="http://www.w3.org/2001/XMLSchema-instance"
```

```xml
            xsi:schemaLocation ="http://maven.apache.org/POM/4.0.0 http://maven.apache.org/xsd/maven-4.0.0.xsd">
    <parent>
        <artifactId>jareny-recommend</artifactId>
        <groupId>com.it.jareny.bigdata</groupId>
        <version>1.0-SNAPSHOT</version>
    </parent>
    <modelVersion>4.0.0</modelVersion>

    <artifactId>dataLoader</artifactId>

    <dependencies>
        <dependency>
            <groupId>com.it.jareny.bigdata</groupId>
            <artifactId>model</artifactId>
            <version>1.0-SNAPSHOT</version>
        </dependency>
        <dependency>
            <groupId>com.it.jareny.bigdata</groupId>
            <artifactId>common</artifactId>
            <version>1.0-SNAPSHOT</version>
        </dependency>
    </dependencies>

</project>
```

2. 维护加载数据配置

新建一个 DataLoadEnum 类，获取加载数据的路径和保存到数据库的集合名称。

```java
package com.jareny.enums;

import lombok.AllArgsConstructor;
import lombok.Getter;

@Getter
@AllArgsConstructor
public enum DataLoadEnum {
    /**
     * 加载数据的配置
     */
    Product("商品数据集",
            DataLoadEnum.class.getResource("/products.csv").getPath(), "Product"),

    UserRating("用户评分数据集",
            DataLoadEnum.class.getResource("/ratings.csv").getPath(), "UserRating");
```

```
    // 数据描述
    private String dateName;
    // 数据路径,当前项目的 resources 文件
    private String filePath;
    // 数据存储集合名称
    private String collectionName;
}
```

3. 加载数据

加载数据的实现步骤:

1) 创建 Spark 和配置 Spark 的资源。

2) 获取商品转换函数。

3) 保存到数据库。

4) 获取商品转换函数。

5) 保存到数据库。

6) 关闭 Spark。

新建一个 DataLoad 类,在 DataLoad 类中实现加载数据的逻辑处理,代码如下。

```
package com.jareny.bigdata;

import com.jareny.bigdata.model.Product;
import com.jareny.bigdata.model.UserRating;
import com.jareny.bigdata.util.ConvertUtil;
import com.jareny.bigdata.util.MongoUtil;
import com.jareny.enums.DataLoadEnum;
import org.apache.spark.SparkConf;
import org.apache.spark.api.java.JavaRDD;
import org.apache.spark.api.java.JavaSparkContext;
import org.apache.spark.api.java.function.Function;
import org.bson.Document;

/**
 * 加载数据
 */
public class DataLoader {

    public static void main(String[] args) {
        // 1.创建 Spark
        SparkConf sparkConf = new SparkConf()
                .setAppName("DataLoader")
                .setMaster("local[*]")
                .set("spark.app.id", "DataLoader")
```

```java
            .set("spark.mongodb.input.uri", "mongodb://192.168.81.111:27017/recommen-
der.recommender")
            .set("spark.mongodb.output.uri", "mongodb://192.168.81.111:27017/recommen-
der.recommender");

    // 2.创建 SparkContext
    JavaSparkContext sparkContext = new JavaSparkContext(sparkConf);

    // 3.获取商品转换函数
    Function<String, Product> productFunction = getProductFunction();
    // 4.保存到数据库
    saveToMongo(sparkContext, DataLoadEnum.Product, productFunction);

    // 5.获取商品转换函数
    Function<String, UserRating> userRatingFunction = getUserRatingFunction();
    // 6.保存到数据库
    saveToMongo(sparkContext, DataLoadEnum.UserRating, userRatingFunction);

    // 7.关闭 Spark
    sparkContext.stop();
}

/**
 * 商品转换函数
 *
 * @return
 */
public static Function<String, Product> getProductFunction() {
    return item -> {
        String[] split = item.split("\\^");
        return new Product(Integer.valueOf(split[0]), split[1].trim(),
                split[4].trim(), split[5].trim(), split[6].trim());
    };
}

/**
 * 用户评分转换函数
 *
 * @return
 */
public static Function<String, UserRating> getUserRatingFunction() {
    return item -> {
        String[] split = item.split(",");
        return new UserRating(Integer.valueOf(split[0]), Integer.valueOf(split[1]),
                Double.valueOf(split[2]), Integer.valueOf(split[3]));
    };
```

```
}
/**
 * 保存到mongodb数据库
 *
 * @param sparkContext spark 配置
 * @param dataLoad     存储的配置
 * @param function     转换函数
 * @param <T>
 */
public static <T> void saveToMongo(JavaSparkContext sparkContext,
                                    DataLoadEnum dataLoad,
                                    Function<String, T> function) {
    // 1.读取文件成 RDD
    JavaRDD<String> stringRdd = sparkContext.textFile(dataLoad.getFilePath());

    // 2.转成 documentRDD
    JavaRDD<Document> documentRdd = ConvertUtil.convertToDocument(stringRdd, function);

    // 3.保存到 Mongodb 数据库
    MongoUtil.saveToMongo(sparkContext, documentRdd, dataLoad.getCollectionName());
}
```

温馨提示：products.csv 和 ratings.csv 文件中存储了练习数据集，可以从 https://gitee.com/jareny/jareny-bigdata.git 上获取。

12.4 离线统计模块

12.4.1 模块介绍

离线统计模块是综合用户所有的历史数据，利用设定的离线统计算法和离线推荐算法周期性地进行结果统计与保存，计算的结果在一定时间周期内是固定不变的，变更的频率取决于算法调度的频率。

离线统计模块主要计算一些可以预先进行统计和计算的指标，为实时计算和前端业务提供数据支撑。

离线推荐服务主要分为统计推荐、基于隐语义模型的协同过滤推荐，以及基于内容和基于 Item-CF 的相似推荐。

12.4.2 模块实现

1. 新建离线统计服务模块并且添加依赖

新建 statisticsRecommender 模块，并且在 statisticsRecommender 模块的 pom.xml 文件中添加依赖，

如下。

```xml
<?xml version="1.0" encoding="UTF-8"?>
<project xmlns="http://maven.apache.org/POM/4.0.0"
      xmlns:xsi="http://www.w3.org/2001/XMLSchema-instance"
      xsi:schemaLocation="http://maven.apache.org/POM/4.0.0 http://maven.apache.org/xsd/maven-4.0.0.xsd">
    <parent>
        <artifactId>jareny-recommend</artifactId>
        <groupId>com.it.jareny.bigdata</groupId>
        <version>1.0-SNAPSHOT</version>
    </parent>
    <modelVersion>4.0.0</modelVersion>

    <artifactId>statisticsRecommender</artifactId>

    <dependencies>
        <dependency>
            <groupId>com.it.jareny.bigdata</groupId>
            <artifactId>model</artifactId>
            <version>1.0-SNAPSHOT</version>
        </dependency>
    </dependencies>

</project>
```

2. 维护加载数据配置

新建 StatisticsEnums 类，并且在 StatisticsEnums 类中维护统计模块数据集合和统计 SQL，代码如下。

```java
package com.jareny.enums;

import lombok.AllArgsConstructor;
import lombok.Getter;

@Getter
@AllArgsConstructor
public enum StatisticsEnums {
    /**
     * 统计数据类型的配置
     */
    UserRating("用户评分数据集", "UserRating",
            "select productId, score, changeDate(timestamp) as yearMonth from ratings"),

    RateMoreProducts("统计热门数据集", "RateMoreProducts",
            "select productId, count(productId) as count from ratings group by productId order by count desc"),
```

```
RateRecentlyProducts("统计近期热门数据集", "RateRecentlyProducts",
        "select productId, count(productId) as count, yearMonth from ratingOfMonth " +
                " group by yearMonth, productId order by yearMonth desc, count desc"),

AverageProducts("统计近期优质商品数据集", "AverageProducts",
        "select productId, avg(score) as avg from ratings group by productId order by avg desc");

// 数据描述
private String dateName;
// 数据存储集合名称
private String collectionName;
// 查询数据集合的 Spark SQL
private String sparkSql;
}
```

3. 离线统计服务模块实现

离线统计的实现步骤如下。

1）定义 Spark 和数据库连接的全局配置。

2）创建 SparkSession 和 JavaSparkContext。

3）加载数据。

4）统计数据。

5）关闭 Spark。

```
package com.jareny.bigdata;

import com.jareny.enums.StatisticsEnums;
import com.mongodb.spark.MongoSpark;
import com.mongodb.spark.config.ReadConfig;
import org.apache.spark.SparkConf;
import org.apache.spark.api.java.JavaSparkContext;
import org.apache.spark.sql.Dataset;
import org.apache.spark.sql.Row;
import org.apache.spark.sql.SparkSession;
import org.apache.spark.sql.api.java.UDF1;
import org.apache.spark.sql.types.DataTypes;
import java.time.Instant;
import java.time.ZoneId;
import java.time.format.DateTimeFormatter;

/**
 * 数据统计推荐
 */
public class StatisticsRecommender {
```

```java
public static void main(String[] args) {
    // 第1步 定义 Spark 和数据库连接的全局配置
    SparkConf sparkConf = new SparkConf().
            setAppName("StatisticsRecommender")
            .setMaster("local[*]")
            .set("spark.app.id", "StatisticsRecommender")
            .set("spark.mongodb.input.uri", "mongodb://192.168.81.111:27017/recommender.recommender")
            .set("spark.mongodb.output.uri", "mongodb://192.168.81.111:27017/recommender.recommender");
    // 第2步 创建 SparkSession,JavaSparkContext
    SparkSession sparkSession = SparkSession.builder().config(sparkConf).getOrCreate();
    // 创建 JavaSparkContext
    JavaSparkContext javaSparkContext = new JavaSparkContext(sparkSession.sparkContext());

    ReadConfig readConfig = ReadConfig.create(javaSparkContext)
            .withOption("collection", StatisticsEnums.UserRating.getCollectionName());

    // 第3步 加载数据
    Dataset<Row> ratingDF = MongoSpark.load(javaSparkContext,readConfig).toDF();
    // 创建 rating 临时表
    ratingDF.createOrReplaceTempView("ratings");

    // 第4步 统计数据
    // (1)统计历史热门数据,并且保存 mongodb 数据库
    saveToMongo(sparkSession, StatisticsEnums.RateMoreProducts);

    // 注册 UDF,将 timestamp 转换为年月格式 yyyyMM,
    UDF1<Integer, Integer> changeDate = (timestamp) -> Integer.parseInt(Instant.ofEpochSecond(timestamp)
            .atZone(ZoneId.systemDefault()).toLocalDate().format(DateTimeFormatter.ofPattern("yyyyMM")));
    // 注册 UDF
    sparkSession.udf().register("changeDate", changeDate, DataTypes.IntegerType);

    // 把原始 rating 数据转换为想要的结构:productId, score, yearMonth
    Dataset<Row> ratingOfYearMonthDF = sparkSession.sql(StatisticsEnums.UserRating.getSparkSql());

    ratingOfYearMonthDF.createOrReplaceTempView("ratingOfMonth");

    // (2)统计近期热门数据,并且保存 mongodb 数据库
    saveToMongo(sparkSession, StatisticsEnums.RateRecentlyProducts);

    // (3)统计优质商品,商品的平均评分,并且保存 mongodb 数据库
```

```
            saveToMongo(sparkSession, StatisticsEnums.AverageProducts);

            // 第 5 步 关闭 Spark
            sparkSession.stop();
            sparkSession.close();
    }

    /**
     * 使用 SparkSql 统计数据,并且保存到 Mongodb 数据库
     *
     * @param sparkSession     spark 会话
     * @param statisticsEnums  统计的类型
     * @param <T>
     */
    public static <T> void saveToMongo(SparkSession sparkSession, StatisticsEnums statisticsEnums) {
        // 根据 StatisticsEnums 类型获取对应的数据集的统计 SparkSql
        Dataset<Row> dataset = sparkSession.sql(statisticsEnums.getSparkSql());
        // 保存到 mogoDB 数据库
        MongoSpark.save(dataset.write().option("collection", statisticsEnums.getCollectionName()).mode("overwrite"));
    }

}
```

12.5 离线推荐模块

12.5.1 模块介绍

项目采用 ALS 作为协同过滤算法，根据数据库中的用户评分表计算离线的用户商品推荐列表，以及商品相似度矩阵。

通过 ALS 计算商品相似度矩阵，该矩阵用于查询当前商品的相似商品并为实时推荐系统服务。

离线计算的 ALS 算法，算法最终会为用户、商品分别生成最终的特征矩阵，分别是表示用户特征矩阵的 U（m×k）矩阵，每个用户由 k 个特征描述。表示物品特征矩阵的 V（n×k）矩阵，每个物品也由 k 个特征描述。

12.5.2 模块实现

1. 新建离线模块并且添加依赖

新建一个 offlineRecommender 模块，并且在 offlineRecommender 模块中添加依赖，如下。

```
<?xml version="1.0" encoding="UTF-8"?>
<project xmlns="http://maven.apache.org/POM/4.0.0"
```

```xml
        xmlns:xsi="http://www.w3.org/2001/XMLSchema-instance"
         xsi:schemaLocation="http://maven.apache.org/POM/4.0.0 http://maven.apache.org/xsd/maven-4.0.0.xsd">
    <parent>
        <artifactId>jareny-recommend</artifactId>
        <groupId>com.it.jareny.bigdata</groupId>
        <version>1.0-SNAPSHOT</version>
    </parent>
    <modelVersion>4.0.0</modelVersion>

    <artifactId>offlineRecommender</artifactId>
    <dependencies>
        <dependency>
            <groupId>com.it.jareny.bigdata</groupId>
            <artifactId>model</artifactId>
            <version>1.0-SNAPSHOT</version>
        </dependency>
        <dependency>
            <groupId>com.it.jareny.bigdata</groupId>
            <artifactId>common</artifactId>
            <version>1.0-SNAPSHOT</version>
        </dependency>
    </dependencies>

</project>
```

2. 离线模块配置维护

新建一个 OfflineRecsEnum 类，然后在 OfflineRecsEnum 类中，维护存储集合和离线推荐的计算逻辑，代码如下。

```java
package com.jareny.enums;

import cn.hutool.json.JSONUtil;
import com.jareny.bigdata.model.ProductRecs;
import com.jareny.bigdata.model.Recommendation;
import com.jareny.bigdata.model.UserRecs;
import lombok.AllArgsConstructor;
import lombok.Getter;
import org.apache.spark.api.java.JavaPairRDD;
import org.apache.spark.api.java.JavaRDD;
import org.apache.spark.mllib.recommendation.MatrixFactorizationModel;
import org.apache.spark.mllib.recommendation.Rating;
import org.bson.Document;
import org.jblas.DoubleMatrix;
import scala.Tuple2;
import java.util.stream.Collectors;
```

```java
import java.util.stream.StreamSupport;

/**
 * 离线推荐枚举类
 */
@Getter
@AllArgsConstructor
public enum OfflineRecsEnum {
    /**
     * 离线推荐统计数据和集合
     */
    UserRecs("获得用户评分矩阵,得到用户的推荐列表", "UserRecs") {
        @Override
        public JavaRDD<Document> getDocumentRdd(MatrixFactorizationModel model, JavaRDD<Integer> userRdd, JavaRDD<Integer> productRdd) {
            // 1.userRDD 和 productRDD 做笛卡儿积
            JavaPairRDD<Integer, Integer> userProducts = userRdd.cartesian(productRdd).mapToPair(pair -> new Tuple2<>(pair._1(), pair._2()));

            // 2.从训练后的数据中,得到用户的推荐商品列表
            JavaRDD<Rating> preRatings = model.predict(userProducts);

            // 3.从预测评分矩阵中提取得到用户推荐列表
            JavaRDD<UserRecs> userRecsRDD = preRatings.filter(rating -> rating.rating() > 0)
                    .mapToPair(rating -> new Tuple2<>(rating.user(), new Tuple2<>(rating.product(), rating.rating())))
                    .groupByKey()
                    .map(tuple -> new UserRecs(tuple._1(),
                            // 通过分数来排序
                            StreamSupport.stream(tuple._2().spliterator(), false)
                                    .sorted((a, b) -> Double.compare(b._2(), a._2()))
                                    .limit(20)
                                    .map(item -> new Recommendation(item._1(), item._2()))
                                    .collect(Collectors.toList())));
            // 4.转换成 mongodb 的数据集合
            return userRecsRDD.map(entityDocument -> Document.parse(JSONUtil.toJsonStr(entityDocument)));
        }
    },

    ProductRecs("利用商品的特征向量,计算商品的相似度列表", "ProductRecs") {
        @Override
        public JavaRDD<Document> getDocumentRdd(MatrixFactorizationModel model, JavaRDD<Integer> userRdd, JavaRDD<Integer> productRdd) {
            // 1.推荐物品的评分进行转换
            JavaPairRDD<Integer, DoubleMatrix> productFeaturesRDD = JavaPairRDD.fromJavaRDD(model.productFeatures().toJavaRDD()
```

```java
                    .map(tuple -> new Tuple2<>(Integer.parseInt(tuple._1().toString()), new
DoubleMatrix(tuple._2()))));

            // 2.两两配对商品,计算余弦相似度
            JavaPairRDD<Integer, Tuple2<Integer, Double>> productSimRDD = productFea-
turesRDD.cartesian(productFeaturesRDD)
                    .filter(tuple -> !tuple._1()._1().equals(tuple._2()._1()))
                    .mapToPair(tuple -> new Tuple2<>(tuple._1()._1(),
                            new Tuple2<>(tuple._2()._1(), consinSim(tuple._1()._2(), tuple._2
()._2()))))
                    .filter(tuple -> tuple._2()._2() > 0.4);

            // 3.获取物品推荐列表,前 20 个
            JavaRDD<ProductRecs> productRecsRDD = productSimRDD.groupByKey()
                    .map(tuple -> new ProductRecs(tuple._1(), StreamSupport.stream(tuple._2
().spliterator(), false)
                            .sorted((a, b) -> Double.compare(b._2(), a._2()))
                            .limit(20)
                            .map(sim -> new Recommendation(sim._1(), sim._2()))
                            .collect(Collectors.toList())));

            // 4.转换成 mongodb 的数据集合
            return productRecsRDD.map(entityDocument -> Document.parse(JSONUtil.toJsonStr
(entityDocument)));
        }
    };

    // 数据描述
    private String dateName;
    // 数据存储集合名称
    private String collectionName;

    /**
     * 根据枚举类型,获取推荐列表数据集
     *
     * @param model
     * @param userRdd
     * @param productRdd
     * @return
     */
    public abstract JavaRDD<Document> getDocumentRdd(MatrixFactorizationModel model,
                                                    JavaRDD<Integer> userRdd,
                                                    JavaRDD<Integer> productRdd);

    // 计算两个商品之间的余弦相似度
    private static double consinSim(DoubleMatrix product1, DoubleMatrix product2) {
```

```
            return product1.dot(product2) / (product1.norm2() * product2.norm2());
    }
}
```

3. 模块的实现

通过 ALS 训练出来的 Model 来计算所有当前用户商品的推荐列表，主要实现步骤如下。

1）创建 SparkConfig 和 SparkSession。
2）读取 mongoDB 中的评分数据。
3）将评分数据变成训练的数据，并且缓存。
4）提取所有用户和商品的数据集。
5）离线推荐计算（核心计算过程）。
6）关闭 Spark。

```
package com.jareny;

import com.jareny.bigdata.model.UserRating;
import com.jareny.bigdata.util.MongoUtil;
import com.jareny.enums.OfflineRecsEnum;
import com.mongodb.spark.MongoSpark;
import com.mongodb.spark.config.ReadConfig;
import org.apache.spark.SparkConf;
import org.apache.spark.api.java.JavaRDD;
import org.apache.spark.api.java.JavaSparkContext;
import org.apache.spark.mllib.recommendation.ALS;
import org.apache.spark.mllib.recommendation.MatrixFactorizationModel;
import org.apache.spark.mllib.recommendation.Rating;
import org.apache.spark.sql.Dataset;
import org.apache.spark.sql.Encoders;
import org.apache.spark.sql.Row;
import org.apache.spark.sql.SparkSession;
import org.bson.Document;

/**
 * 离线推荐实现
 */
public class OfflineRecommender {

    public static void main(String[] args) {
        // 第1步 创建 SparkConfig 和 SparkSession
        SparkConf sparkConf = new SparkConf()
                .setMaster("local[*]")
                .setAppName("OfflineRecommender")
                .set("spark.mongodb.input.uri", "mongodb://192.168.81.111:27017/recommender.recommender")
```

```java
            .set("spark.mongodb.output.uri", "mongodb://192.168.81.111:27017/recommender.recommender")
            .set("spark.mongodb.output.database", "recommender");

    // 创建 spark session
    SparkSession spark = SparkSession.builder().config(sparkConf).getOrCreate();

    JavaSparkContext javaSparkContext = new JavaSparkContext(spark.sparkContext());

    // 第 2 步 读取 mongoDB 中的评分数据
    ReadConfig readConfig = ReadConfig.create(javaSparkContext)
            .withOption("collection", "UserRating");

    Dataset<Row> ratingDF = MongoSpark.load(javaSparkContext, readConfig).toDF();

    // 第 3 步 将评分数据变成训练的数据,并且缓存
    JavaRDD<Rating> ratingRDD = ratingDF.as(Encoders.bean(UserRating.class))
            .javaRDD()
            .map(userRating -> new Rating(userRating.getUserId(), userRating.getProductId(), userRating.getScore()))
            .cache();

    // 第 4 步 提取所有用户和商品的数据集
    JavaRDD<Integer> userRDD = ratingRDD.map(Rating::user).distinct();
    JavaRDD<Integer> productRDD = ratingRDD.map(Rating::product).distinct();

    // 第 5 步 离线推荐计算(核心计算过程)
    // 5.1 定义训练的参数(因子),训练隐语义模型
    MatrixFactorizationModel model = ALS.train(JavaRDD.toRDD(ratingRDD), 5, 10, 0.01, 1);

    // 5.2 获得用户评分矩阵,得到用户的推荐列表,并且保存到 mongodb 数据库
    saveToMongo(javaSparkContext, model, userRDD, productRDD, OfflineRecsEnum.UserRecs);

    // 5.3 利用商品的特征向量,计算商品的相似度列表,并且保存到 mongodb 数据库
    saveToMongo(javaSparkContext, model, userRDD, productRDD, OfflineRecsEnum.ProductRecs);

    // 第 6 步 关闭 Spark
    spark.stop();
    spark.close();
}

/**
 * 保存数据到 MongoDB 数据库
 *
 * @param javaSparkContext Spark 上下文
 * @param model            模型
```

```
 * @param userRdd          用户数据集
 * @param productRdd       商品数据集
 * @param offlineRecsEnum  离线统计(配置)枚举类维护
 * @return
 */
private static void saveToMongo(JavaSparkContext javaSparkContext,
                                MatrixFactorizationModel model,
                                JavaRDD<Integer> userRdd,
                                JavaRDD<Integer> productRdd,
                                OfflineRecsEnum offlineRecsEnum) {
    //1.根据枚举类型获取推荐的列表
    JavaRDD<Document> documentRdd = offlineRecsEnum.getDocumentRdd(model, userRdd, productRdd);
    //2.保存到 MongoDB 数据库
    MongoUtil.saveToMongo(javaSparkContext, documentRdd, offlineRecsEnum.getCollectionName());
}
```

12.5.3 离线推荐模型评估

在模型训练的过程中，直接给定了隐语义模型的 rank、iterations、lambda 三个参数。对于模型，这并不一定是最优的参数选取，所以需要对模型进行评估。通常的做法是计算均方根误差（RMSE），考察预测评分与实际评分之间的误差。

有了 RMSE，我们就可以通过多次调整参数值，来选取 RMSE 最小的一组作为模型的优化选择。

1. 定义离线推荐模型评估的训练参数

新建一个 AlsTrainFactors 类，使用 AlsTrainFactors 来维护离线推荐模型的训练参数。

```
package com.jareny.bigdata;

import com.google.common.collect.ImmutableSet;
import com.google.common.collect.Sets;
import lombok.AllArgsConstructor;
import lombok.Data;
import lombok.NoArgsConstructor;
import java.util.ArrayList;
import java.util.List;
import java.util.Set;

/**
 * 训练参数
 */
```

```java
@Data
@NoArgsConstructor
@AllArgsConstructor
public class AlsTrainFactors {
    // 隐藏因子的个数
    private int rank;

    // 迭代次数
    private int iters;

    // 正则项的惩罚系数
    private double lambdas;

    // 隐藏因子的个数列表
    private static final ImmutableSet<String> rankList = ImmutableSet.of("5", "10", "15");
    // 迭代次数列表
    private static final ImmutableSet<String> itersList = ImmutableSet.of("8", "10");
    // 正则项的惩罚系数列表
    private static final ImmutableSet<String> lambdaList = ImmutableSet.of("1", "0.1", "0.01");

    /**
     * 获取训练因子
     *
     * @return
     */
    public static List<AlsTrainFactors> getAlsTrainList() {
        Set<List<String>> set = Sets.cartesianProduct(rankList, itersList, lambdaList);
        List<AlsTrainFactors> alsTrainFactorsList = new ArrayList<>();
        for (List<String> list : set) {
            AlsTrainFactors alsTrainFactors = new AlsTrainFactors();
            alsTrainFactors.setRank(Integer.parseInt(list.get(0)));
            alsTrainFactors.setIters(Integer.parseInt(list.get(1)));
            alsTrainFactors.setLambdas(Double.parseDouble(list.get(2)));
            alsTrainFactorsList.add(alsTrainFactors);
        }
        return alsTrainFactorsList;
    }
}
```

2. 定义离线推荐模型评估的训练结果

新建一个 AdjustAlsVariance 类，使用 AdjustAlsVariance 来维护离线推荐模型的训练参数。

```java
package com.jareny.bigdata;

import org.apache.spark.mllib.recommendation.MatrixFactorizationModel;
```

```java
/**
 * ALS 训练集结果
 */
public class AdjustAlsVariance {
    // 训练因子
    private AlsTrainFactors alsTrain;

    // 模型
    private MatrixFactorizationModel model;

    // 方差
    private Double variance;

    public AdjustAlsVariance() {
    }

     public AdjustAlsVariance (AlsTrainFactors alsTrain, MatrixFactorizationModel model, Double variance) {
        this.alsTrain = alsTrain;
        this.model = model;
        this.variance = variance;
    }

    public AlsTrainFactors getAlsTrain() {
        return alsTrain;
    }

    public void setAlsTrain(AlsTrainFactors alsTrain) {
        this.alsTrain = alsTrain;
    }

    public MatrixFactorizationModel getModel() {
        return model;
    }

    public void setModel(MatrixFactorizationModel model) {
        this.model = model;
    }

    public Double getVariance() {
        return variance;
    }

    public void setVariance(Double variance) {
        this.variance = variance;
    }
}
```

3. 离线推荐模型的评估实现

离线推荐模型的评估实现步骤如下。

1) 创建 SparkConfig 和 SparkSession。
2) 引读取 mongoDB 中的评分数据，并且存储到缓存。
3) 将评分数据变成训练的数据，并且缓存。
4) 定义拆分比例，例如 70%用于训练，15%用于测试，15%用于验证。
5) 获取训练的优化模型。
6) 训练后获取最好的模型，根据最好的模型及训练集 testData 来获取此方差。
7) 获取评估模型是否优化。
8) 根据平均值来计算旧的方差值。
9) 通过模型，计算数据的拟合度。
10) 关闭 Spark。

```java
package com.jareny.bigdata;

import cn.hutool.core.collection.CollUtil;
import com.mongodb.spark.MongoSpark;
import lombok.extern.slf4j.Slf4j;
import org.apache.spark.SparkConf;
import org.apache.spark.api.java.JavaPairRDD;
import org.apache.spark.api.java.JavaRDD;
import org.apache.spark.api.java.JavaSparkContext;
import org.apache.spark.api.java.function.DoubleFunction;
import org.apache.spark.api.java.function.Function;
import org.apache.spark.api.java.function.Function2;
import org.apache.spark.mllib.recommendation.ALS;
import org.apache.spark.mllib.recommendation.MatrixFactorizationModel;
import org.apache.spark.mllib.recommendation.Rating;
import org.apache.spark.sql.Dataset;
import org.apache.spark.sql.Row;
import org.apache.spark.sql.SparkSession;
import scala.Tuple2;
import java.util.ArrayList;
import java.util.Comparator;
import java.util.List;

/**
 * 离线推荐优化训练
 */
@Slf4j
public class ALSTrainer {

    public static void main(String[] args) {
```

第 12 章 电商推荐系统实战

```java
// 第 1 步 创建 SparkConfig 和 SparkSession
SparkConf sparkConf = new SparkConf()
        .setMaster("local[*]")
        .setAppName("OfflineRecommender")
        .set("spark.mongodb.input.uri", "mongodb://192.168.81.111:27017/recommender.Rating")
        .set("spark.mongodb.output.uri", "mongodb://192.168.81.111:27017/recommender.Rating")
        .set("spark.mongodb.output.database", "recommender");

// 创建 spark session
SparkSession spark = SparkSession.builder().config(sparkConf).getOrCreate();

JavaSparkContext javaSparkContext = new JavaSparkContext(spark.sparkContext());

// 第 2 步 引读取 mongoDB 中的评分数据,并且存储到缓存
Dataset<Row> ratingDF = MongoSpark.load(javaSparkContext).toDF();

// 第 3 步 将评分数据变成训练的数据,并且缓存
JavaRDD<Rating> ratingRDD = ratingDF
        .toJavaRDD()
        .map(row -> {
            int userId = row.getAs("userId");
            int productId = row.getAs("productId");
            double score = row.getAs("score");
            return new Rating(userId, productId, score);
        }).cache();

// 第 4 步 定义拆分比例,例如 70%用于训练,15%用于测试,15%用于验证
double[] fractions = {0.7, 0.15, 0.15};

// 将 JavaRDD 拆分成多个部分
JavaRDD<Rating>[] datasets = ratingRDD.randomSplit(fractions);

// 分别处理训练、测试和验证数据集
JavaRDD<Rating> trainingData = datasets[0];
JavaRDD<Rating> testData = datasets[1];
JavaRDD<Rating> validationData = datasets[2];

// 第 5 步 获取训练的优化模型
MatrixFactorizationModel model = getAdjustMatrixFactorizationModel(trainingData, validationData);

// 第 6 步 训练后获取最好的模型,根据最好的模型及训练集 testData 来获取此方差
double testDataRnse = getVariance(model, testData, testData.count());
```

```java
        // 第 7 步 获取评估模型是否优化
        // 获取测试数据中,分数的平均值
        double meanRating = testData.union(validationData).mapToDouble(new DoubleFunction
<Rating>() {
            @Override
            public double call(Rating t) {
                return t.rating();
            }
        }).mean();

        // 第 8 步 根据平均值来计算旧的方差值
        double baseLineRnse = Math.sqrt(testData.mapToDouble(new DoubleFunction<Rating>() {
            @Override
            public double call(Rating t) {
                return (meanRating - t.rating()) * (meanRating - t.rating());
            }
        }).mean());

        // 第 9 步 通过模型,计算数据的拟合度
        double improvent = (baseLineRnse - testDataRnse) / baseLineRnse * 100;
        // 打印模型数据拟合度
        log.info("model improves is {}", improvent);

        // 第 10 步 关闭 Spark
        javaSparkContext.stop();
    }

    /**
     * 获取调整后的模型
     *
     * @return
     */
    public static MatrixFactorizationModel getAdjustMatrixFactorizationModel(JavaRDD<Rating>
trainingData, JavaRDD<Rating> validationData) {
        // 训练的参数模型和结果列表
        List<AdjustAlsVariance> list = new ArrayList<>();

        // 获取训练参数列表
        List<AlsTrainFactors> alsTrainList = AlsTrainFactors.getAlsTrainList();

        // 循环训练模型
        for (AlsTrainFactors alsTrain : alsTrainList) {
            // 训练参数来训练模型
            MatrixFactorizationModel model = ALS.train(JavaRDD.toRDD(trainingData),
                    alsTrain.getRank(), alsTrain.getIters(), alsTrain.getLambdas());
            // 通过校验集 validationData 获取方差,以便查看此模型的好坏
```

```java
            double variance = getVariance(model, validationData, validationData.count());

            // 组装训练参数
            AdjustAlsVariance adjustAlsVariance = new AdjustAlsVariance();
            adjustAlsVariance.setAlsTrain(alsTrain);
            adjustAlsVariance.setModel(model);
            adjustAlsVariance.setVariance(variance);

            // 收集训练后的数据因子、方差、模型
            list.add(adjustAlsVariance);
        }
        if (CollUtil.isEmpty(list)) {
            // 训练模型,默认训练次数
            return ALS.train(JavaRDD.toRDD(trainingData), 10, 5, 0.1);
        }
        // 得到更优的训练参数
        AdjustAlsVariance adjustAlsVariance = list.stream()
                .sorted(Comparator.comparing(AdjustAlsVariance::getVariance).reversed()).findFirst().get();
        // 返回训练最优的模型
        return adjustAlsVariance.getModel();
    }

    /**
     * 调整 ALS 的参数,根据方差决定
     *
     * @param model
     * @param predictionData
     * @param n
     * @return
     */
    public static double getVariance(MatrixFactorizationModel model, JavaRDD<Rating> predictionData, long n) {
        // 1.提取用户和商品的信息
        JavaRDD<Tuple2<Object, Object>> userProducts = predictionData.map(new Function<Rating, Tuple2<Object, Object>>() {
            @Override
            public Tuple2<Object, Object> call(Rating r) {
                return new Tuple2<Object, Object>(r.user(), r.product());
            }
        });

        // 2.通过模型对数据进行预测
        JavaPairRDD<Tuple2<Integer, Integer>, Double> prediction = JavaPairRDD.fromJavaRDD
(model.predict(JavaRDD.toRDD(userProducts))
                .toJavaRDD().map(new Function<Rating, Tuple2<Tuple2<Integer, Integer>, Double>>() {
```

```
            @Override
            public Tuple2<Tuple2<Integer, Integer>, Double> call(Rating r) {
                return new Tuple2<Tuple2<Integer, Integer>, Double>(new Tuple2<Integer,
Integer>(r.user(), r.product()), r.rating());
            }
        }));

        // 3.预测值和原值内连接
        JavaRDD<Tuple2<Double, Double>> ratesAndPreds = JavaPairRDD.fromJavaRDD(
                predictionData.map(new Function<Rating, Tuple2<Tuple2<Integer, Integer>,
Double>>() {
            @Override
            public Tuple2<Tuple2<Integer, Integer>, Double> call(Rating r) {
                return new Tuple2<Tuple2<Integer, Integer>, Double>(new Tuple2<Integer, Integer>(r.user(), r.product()), r.rating());
            }
        })).join(prediction).values();

        // 4.计算方差并返回结果
        Double dVar = ratesAndPreds.map(new Function<Tuple2<Double, Double>, Double>() {
            @Override
            public Double call(Tuple2<Double, Double> v1) {
                return (v1._1 - v1._2) * (v1._1 - v1._2);
            }
        }).reduce(new Function2<Double, Double, Double>() {
            @Override
            public Double call(Double v1, Double v2) {
                return v1 + v2;
            }
        });
        // 5.返回方差结果
        return Math.sqrt(dVar / n);
    }
}
```

12.6 在线推荐模块

12.6.1 模块介绍

实时计算与离线计算应用于推荐系统上最大的不同，在于实时计算推荐结果应该反映最近一段时间用户的偏好，而离线计算推荐结果则是根据用户从第一次评分开始的所有评分记录来计算用户总体的偏好。

用户对物品的偏好随着时间的推移总是会改变的。比如一个用户 u 在某时刻对商品 p 给予了极高的评分，那么在近期 u 极有可能很喜欢与商品 p 类似的其他商品，而如果用户 u 在某时刻对商品 q 给予了极低的评分，那么在近期 u 极有可能不喜欢与商品 q 类似的其他商品。所以对于实时推荐，当用户对一个商品进行了评价后，用户会希望推荐结果基于最近这几次评分进行一定的更新，使得推荐结果匹配用户近期的偏好，满足用户近期的口味。

如果实时推荐继续采用离线推荐中的 ALS 算法，由于算法运行时间很长，不具有实时得到新的推荐结果的能力，并且由于算法本身使用的是评分表，用户本次评分后只更新了总评分表中的一项，使得算法运行后的推荐结果与用户本次评分之前的推荐结果没有多少差别，从而给用户一种推荐结果一直没变化的感觉，很影响用户体验。

另外，在实时推荐中由于时间性能上要满足实时或者准实时的要求，所以算法的计算量不能太大，避免复杂、过多的计算造成用户体验的下降。鉴于此，推荐精度往往不会很高。实时推荐系统更关心推荐结果的动态变化能力，只要更新推荐结果的理由合理即可，至于推荐的精度要求则可以适当放宽。

对于实时推荐算法，其核心在于用户完成本次评分或最近几次评分后，系统能够迅速并显著地更新推荐结果，以满足响应时间上的实时或准实时要求。这样的算法能够确保用户在获得即时反馈的同时，也能获得更加准确和符合当前偏好的推荐内容。

12.6.2 需求分析

当用户 u 对商品 p 进行了评分，将触发一次对 u 的推荐结果的更新。由于用户 u 对商品 p 评分，对于用户 u 来说，与 p 最相似的商品之间的推荐强度将发生变化，所以选取与商品 p 最相似的 K 个商品作为候选商品。

每个候选商品将"推荐优先级"这一权重作为衡量这个商品被推荐给用户 u 的优先级。

这些商品将根据用户 u 最近的若干评分计算出各自对用户 u 的推荐优先级，然后与上次对用户 u 的实时推荐结果进行基于推荐优先级的合并、替换得到更新后的推荐结果。

首先，获取用户 u 按时间顺序最近的 K 个评分，记为 RK。

获取商品 p 的最相似的 K 个商品集合，记为 S。

然后，对于每个商品 q，我们需要大数据技术来计算其在电商推荐系统 S 的推荐优先级，这个过程涉及对大量用户行为数据的分析，以了解用户的购买历史、浏览记录、搜索行为等，从而为每个商品生成一个推荐优先级排名，计算公式如下：

$$\sum_{uq} = \frac{\sum_{r \in RK} \text{sim}(q,r) \times R_r}{\text{sim\_sum}} + \lg \max\{\text{incount}, 1\} - \lg \max\{\text{recount}, 1\}$$

式中：

R_r 表示用户 u 对商品 r 的评分。

sim(q,r) 表示商品 q 和商品 r 的相似度，设定最小相似度为 0.6，当商品 q 和商品 r 相似度低于 0.6 的阈值，则视为两者不相关并忽略。

sim_sum 表示 q 与 RK 中商品相似度大于最小阈值的个数。

incount 表示 RK 中与商品 q 相似，且本身评分较高（>=）的商品个数。

recount 表示 RK 中与商品 q 相似，且本身评分较低（<=）的商品个数。

公式的意义如下。

首先对于每一个候选商品 q，从 u 最近的 K 个评分中，找出与 q 相似度较高（>=）的 u 已评分商品，对于这些商品中的每个商品 r，将 r 与 q 的相似度乘以用户 u 对 r 的评分，将这些乘积算平均分，作为用户 u 对商品 q 的评分预测。

$$\frac{\sum_{r \in RK} sim(q,r) \times R_r}{sim\_sum}$$

然后，将最近的 K 个评分中与商品 q 相似的，且本身评分较高（>=3）的商品个数记为 incount，计算 lgmax{incount,1} 作为商品 q 的"增强因子"，意义在于商品 q 与 u 的最近 K 个评分中的 n 个高评分（>=3）商品相似，则商品 q 的优先级被增加 lgmax{incount,1}，如果商品 q 与 u 的最近 K 个评分中相似的高评分商品越多，也就是说 n 越大，则商品 q 更应该被推荐，所以推荐优先级被增强的幅度较大，如果商品 q 与 u 的最近 K 个评分中相似的高评分商品越少，也就是 n 越小，则推荐优先级被增强的幅度较小。

而后，将最近的 K 个评分中与商品 q 相似的，且本身评分较低（<3）的商品个数记为 recount，计算 lgmax{recount,1} 作为商品 q 的"削弱因子"，意义在于商品 q 与 u 的最近 K 个评分中的 n 个低评分（<3）商品相似，则商品 q 的优选级被削减 lgmax{recount,1}。如果商品 q 与 u 的最近 K 个评分中相似的低评分商品越多，也就是说 n 越大，则商品 q 不应该被推荐，所以推荐优先级被减弱的幅度较大；如果商品 q 与 u 的最近 K 个评分中相似的低评分商品越少，也就是 n 越小，则推荐优先级被减弱的幅度较小。

最后，将增强因子增加到上述的预测评分中，并减去削弱因子，得到最终的 q 商品对于 u 的推荐优先级。在计算完每个候选商品 q 的 Euq 后，将生成一组<商品 q 的 ID，q 的推荐优先级>的列表 updateList。

$$updateLits = \bigcup_{q \in s} \{qID, E_{uq}\}$$

而在本次用户 u 实时途径之前的上一次推荐结果 Rec 也是一组<商品 m，m 的推荐优先级>的列表，其大小也为 K。

$$Rec = \bigcup_{m \in Rec} \{mID, E_{nm}\}, len(Rec) = K$$

接下来，将 updatad_S 与本次 u 实时推荐之前的上一次实时推荐结果 Rec 进行合并、替换，形成新的推荐结果 NewRec。

$$NewRec = topK(i \in Rec \cup updateLits, cmp = E_m)$$

式中，i 表示 updated_S 与 Rec 的商品集合中的每个商品；topK 是一个函数，表示从 Rec 与 updataed_S 中选择出最大的 K 个商品。最终，NewRec 即为经过用户 u 对商品 p 评分后触发的实时推荐得到的最新推荐结果。

实时推荐算法流程如下。

1) 用户 u 对商品 p 进行了评分，触发了实时推荐的一次计算。
2) 选出与商品 p 最相似的 K 个商品作为集合 S。
3) 获取用户 u 最近时间内的 K 条评分，包含本次评分，作为集合 RK。

4）计算商品的推荐优先级，产生 qId 集合，update_S。

将 updated_S 与上次对用户 u 的推荐结果 Rec 利用公式进行合并，产生新的推荐结果 NewRec，作为最终输出。

12.6.3 模块实现

1. 新建实时推荐模块并添加依赖

新建 onlineRecommender 模块，并且在 onlineRecommender 模块的 pom.xml 文中添加依赖，代码如下。

```xml
<?xml version="1.0" encoding="UTF-8"?>
<project xmlns="http://maven.apache.org/POM/4.0.0"
      xmlns:xsi="http://www.w3.org/2001/XMLSchema-instance"
      xsi:schemaLocation="http://maven.apache.org/POM/4.0.0 http://maven.apache.org/xsd/maven-4.0.0.xsd">
    <parent>
        <artifactId>jareny-recommend</artifactId>
        <groupId>com.it.jareny.bigdata</groupId>
        <version>1.0-SNAPSHOT</version>
    </parent>
    <modelVersion>4.0.0</modelVersion>

    <artifactId>onlineRecommender</artifactId>

    <dependencies>
        <dependency>
            <groupId>com.it.jareny.bigdata</groupId>
            <artifactId>model</artifactId>
            <version>1.0-SNAPSHOT</version>
        </dependency>
        <!-- 导入 jedis-->
        <!-- https://mvnrepository.com/artifact/redis.clients/jedis -->
        <dependency>
            <groupId>redis.clients</groupId>
            <artifactId>jedis</artifactId>
            <version>3.5.2</version>
        </dependency>
    </dependencies>

</project>
```

2. 数据库访问工具

新建一个 MongoDBUtil 类，用于实时推荐的时候，访问 MongoDB 数据库。代码如下。

```
package com.jareny.bigdata.recommender;

import com.mongodb.client.MongoClient;
```

```java
import com.mongodb.client.MongoClients;
import com.mongodb.client.MongoCollection;
import com.mongodb.client.MongoDatabase;
import org.bson.Document;

public class MongoDBUtil {
    // mongoDB 连接
    private static final String MONGODB_URI = "mongodb://192.168.81.111:27017/recommender";
    // mongoDB 库名
    private static final String DATABASE_NAME = "recommender";

    /**
     * 获取连接
     *
     * @return
     */
    public static MongoDatabase getMongoDatabase() {
        // 连接 MongoDB 数据库
        MongoClient mongoClient = MongoClients.create(MONGODB_URI);
        // 获取 MongoDatabase 对象
        return mongoClient.getDatabase(DATABASE_NAME);
    }


    /**
     * 获取连接
     *
     * @return
     */
    public static MongoDatabase getMongoDatabase(String DatabaseName) {
        // 连接 MongoDB 数据库
        MongoClient mongoClient = MongoClients.create(MONGODB_URI);
        // 获取 MongoDatabase 对象
        return mongoClient.getDatabase(DatabaseName);
    }


    /**
     * 获取文档
     *
     * @return
     */
    public static MongoCollection<Document> getMongoDocument(String collectionName) {
        // 连接 MongoDB 数据库
        MongoDatabase mongoDatabase = getMongoDatabase();
        // 获取 MongoDatabase 对象
        return mongoDatabase.getCollection(collectionName);
    }
}
```

第12章 电商推荐系统实战

3. 在线推荐实现

在线实时推荐算法过程，首先，获取 userId 最近 K 次评分；然后获取 productId 最相似的 K 个商品；计算候选商品的推荐优先级；最后更新对 userId 的实时推荐结果。

实时推荐实现步骤如下。

1）创建 SparkSession、JavaStreamingContext。
2）加载的是离线统计的相似度矩阵，并且广播出去。
3）创建一个 DStream，从 Kafka 接收数据。
4）定义评分流的处理流程（核心算法流程）。
5）启动 SparkStreaming。

实时推荐案例代码如下。

```java
package com.jareny.bigdata.recommender;

import cn.hutool.core.bean.BeanUtil;
import com.jareny.bigdata.model.ProductRecs;
import com.jareny.bigdata.model.Recommendation;
import com.mongodb.client.MongoCollection;
import com.mongodb.client.MongoDatabase;
import com.mongodb.spark.MongoSpark;
import lombok.extern.slf4j.Slf4j;
import org.apache.kafka.clients.consumer.ConsumerRecord;
import org.apache.kafka.common.serialization.StringDeserializer;
import org.apache.spark.SparkConf;
import org.apache.spark.api.java.JavaRDD;
import org.apache.spark.api.java.JavaSparkContext;
import org.apache.spark.api.java.function.Function;
import org.apache.spark.api.java.function.VoidFunction;
import org.apache.spark.broadcast.Broadcast;
import org.apache.spark.sql.Dataset;
import org.apache.spark.sql.Encoders;
import org.apache.spark.sql.Row;
import org.apache.spark.sql.SparkSession;
import org.apache.spark.streaming.Duration;
import org.apache.spark.streaming.api.java.JavaDStream;
import org.apache.spark.streaming.api.java.JavaInputDStream;
import org.apache.spark.streaming.api.java.JavaStreamingContext;
import org.apache.spark.streaming.kafka010.ConsumerStrategies;
import org.apache.spark.streaming.kafka010.KafkaUtils;
import org.apache.spark.streaming.kafka010.LocationStrategies;
import org.bson.Document;
import redis.clients.jedis.Jedis;
import scala.Tuple4;

import java.util.*;
```

```java
import java.util.stream.Collectors;

/**
 * 在线推荐模型
 */
@Slf4j
public class OnlineRecommender {

    public static void main(String[] args) throws Exception {
        // 第 1 步 创建 SparkSession、JavaStreamingContext
        // 1.1.创建 spark conf
        SparkConf sparkConf = new SparkConf()
                .setMaster("local[*]")
                .setAppName("OnlineRecommender")
                .set("spark.mongodb.input.uri", "mongodb://192.168.81.111:27017/recommender.ProductRecs")
                .set("spark.mongodb.output.uri", "mongodb://192.168.81.111:27017/recommender.ProductRecs")
                .set("spark.mongodb.output.database", "recommender");

        // 1.2.创建 JavaStreamingContext
        JavaStreamingContext javaStreamingContext = new JavaStreamingContext(sparkConf, new Duration(2000));
        // 1.3.创建 SparkSession
        SparkSession spark = SparkSession.builder().config(sparkConf).getOrCreate();
        // 1.4.创建 JavaSparkContext
        JavaSparkContext javaSparkContext = new JavaSparkContext(spark.sparkContext());

        // 第 2 步 加载的是离线统计的相似度矩阵,并且广播出去
        Dataset<Row> dataset = MongoSpark.load(javaSparkContext).toDF();

        // 2.1.将数据转换成 map 形式,为了后续查询相似度方便
        Map<Integer, Map<Integer, Double>> simProductsMatrixMap = transformProductsMatrixMap(dataset);

        // 2.2.定义广播变量
        Broadcast<Map<Integer, Map<Integer, Double>>> simProcutsMatrix = javaStreamingContext.sparkContext().broadcast(simProductsMatrixMap);

        // 第 3 步 创建一个 DStream,从 kafka 接收数据
        // 3.1.创建 kafka 配置参数
        Map<String, Object> kafkaParams = new HashMap<>();
        kafkaParams.put("bootstrap.servers", "192.168.81.111:9092");
        kafkaParams.put("key.deserializer", StringDeserializer.class);
        kafkaParams.put("value.deserializer", StringDeserializer.class);
        kafkaParams.put("group.id", "recommender");
```

```java
        kafkaParams.put("auto.offset.reset", "latest");
        kafkaParams.put("enable.auto.commit", false);

        // 3.2.订阅 Kafka 的主题为 recommender
        Collection<String> topics = Arrays.asList("recommender");

        // 3.3.从 kafka 接收数据
        JavaInputDStream<ConsumerRecord<String, String>> kafkaStream = KafkaUtils.createDirectStream(
                javaStreamingContext,
                LocationStrategies.PreferConsistent(),
                ConsumerStrategies.Subscribe(topics, kafkaParams)
        );

        // 3.4.对 kafkaStream 进行处理,产生评分流,userId |productId |score |timestamp
        JavaDStream<Tuple4<Integer, Integer, Double, Integer>> ratingStream = kafkaStream
                .map(new Function<ConsumerRecord<String, String>, Tuple4<Integer, Integer, Double, Integer>>() {
                    @Override
                    public Tuple4<Integer, Integer, Double, Integer> call(ConsumerRecord<String, String> msg) throws Exception {
                        String[] attr = msg.value().split("\\|");
                        return new Tuple4<>(Integer.parseInt(attr[0]), Integer.parseInt(attr[1]), Double.parseDouble(attr[2]), Integer.parseInt(attr[3]));
                    }
                });

        // 第 4 步 定义评分流的处理流程(核心算法流程)
        ratingStream.foreachRDD(new VoidFunction<JavaRDD<Tuple4<Integer, Integer, Double, Integer>>>() {
            @Override
            public void call(JavaRDD<Tuple4<Integer, Integer, Double, Integer>> rdds) {
                rdds.foreachPartition(new VoidFunction<Iterator<Tuple4<Integer, Integer, Double, Integer>>>() {
                    @Override
                    public void call(Iterator<Tuple4<Integer, Integer, Double, Integer>> it) throws Exception {
                        // 4.保存实时推荐列表
                        savaStreamRecs(it, simProcutsMatrix);
                    }
                });
            }
        });

        // 第 5 步 启动 SparkStreaming
        javaStreamingContext.start();
```

```java
            javaStreamingContext.awaitTermination();
    }

    /**
     * 保存实时推荐列表
     *
     * @param it
     * @param simProcutsMatrix
     */
    private static void savaStreamRecs(Iterator<Tuple4<Integer, Integer, Double, Integer>> it,
                        Broadcast<Map<Integer, Map<Integer, Double>>> simProcutsMatrix) {
        while (it.hasNext()) {
            Tuple4<Integer, Integer, Double, Integer> tuple = it.next();
            int userId = tuple._1();
            int productId = tuple._2();
            double score = tuple._3();
            int timestamp = tuple._4();

            // 4.1 从 redis 里取出当前用户的最近评分,保存成一个数组 Array[(productId, score)]
            List<Recommendation> userRecentlyRatings = getUserRecentlyRatings(20, userId);

            // 4.2 从相似度矩阵中获取当前商品最相似的商品列表,作为备选列表,保存成一个数组 Array[productId]
            List<Integer> candidateProducts = getTopSimProducts(20, productId, userId, simProcutsMatrix);

            // 4.3 计算每个备选商品的推荐优先级,得到当前用户的实时推荐列表,保存成 Array[(productId, score)]
            List<Recommendation> streamRecs = computeProductScore(candidateProducts, userRecentlyRatings, simProcutsMatrix);

            // 4.4 把推荐列表保存到 mongodb
            saveDataToMongoDB(userId, streamRecs);
        }
    }

    // 将 Dataset 转成 Map 形式
    private static Map<Integer, Map<Integer, Double>> transformProductsMatrixMap(Dataset<Row> simProductsMatrix) {
        // 转换成物品推荐类
        JavaRDD<ProductRecs> productRecsRDD = simProductsMatrix.as(Encoders.bean(ProductRecs.class))
                .toJavaRDD()
                .map(productRecs -> BeanUtil.copyProperties(productRecs, ProductRecs.class));

        // 转成 map 形式
```

```java
        Map<Integer, Map<Integer, Double>> simProductsMatrixMap = productRecsRDD.collect().stream()
                .collect(Collectors.toMap(ProductRecs::getProductId,
                        // Value 值
                        productRecs -> productRecs.getRecs().stream()
                                .collect(Collectors.toMap(Recommendation::getProductId, Recommendation::getScore))));

        return simProductsMatrixMap;
    }

    // 4.1 从 redis 中获取用户的评分数据
    public static List<Recommendation> getUserRecentlyRatings(int num, int userId) {

        // (1) 从 redis 中获取用户的评分数据,
        // 键名为 uid:USERID,值格式是 PRODUCTID:SCORE
        List<String> ratings = getJedis("192.168.81.111", 6379)
                .lrange("userId:" + userId, 0, num);

        // (2)将评分的数据转成评分列表
        List<Recommendation> ratingPairs = ratings.stream().map(item -> {
            String[] attrs = item.split("\\:");
            return new Recommendation(Integer.parseInt(attrs[0].trim()),
                    Double.parseDouble(attrs[1].trim()));
        }).collect(Collectors.toList());
        return ratingPairs;
    }

    //  4.2 从相似度矩阵中获取当前商品的相似列表,
    //  并过滤掉用户已经评分过的,作为备选列表
    public static List<Integer> getTopSimProducts(int num, int productId, int userId, Broadcast<Map<Integer, Map<Integer, Double>>> simProcutsMatrix) {
        // (1) 从广播变量相似度矩阵中拿到当前商品的相似度列表
        Map<Integer, Map<Integer, Double>> simProducts = simProcutsMatrix.value();
        List<Integer> allSimProducts = new ArrayList<>(simProducts.get(productId).keySet());

        // (2) 获取用户已经评分过的商品,过滤掉,排序输出
        MongoDatabase database = MongoDBUtil.getMongoDatabase("recommender");
        MongoCollection<Document> ratingCollection = database.getCollection("UserRating");
        Document query = new Document("userId", userId);
        List<Document> ratingDocs = ratingCollection.find(query).into(new ArrayList<>());
        List<Integer> ratingExist = new ArrayList<>();
        for (Document doc : ratingDocs) {
            ratingExist.add(doc.getInteger("productId"));
        }
```

```java
        // (3) 从所有的相似商品中进行过滤
        List<Integer> filteredSimProducts = allSimProducts.stream()
                .filter(x -> ! ratingExist.contains(x))
                .sorted(Comparator.comparingDouble(x -> simProducts.get(x).get(productId)))
                .limit(num)
                .collect(Collectors.toList());

        return filteredSimProducts;
    }

// 4.3 计算每个备选商品的推荐得分,得到当前用户的实时推荐列表
public static List<Recommendation> computeProductScore(List<Integer> candidateProducts,
                    List<Recommendation> userRecentlyRatings,
                    Broadcast<Map<Integer, Map<Integer, Double>>> simProcutsMatrix) {
    // (1) 定义一个长度可变数组 ArrayBuffer,用于保存每一个备选商品的基础得分,(productId, score)
    Map<Integer, Map<Integer, Double>> simProducts = simProcutsMatrix.value();

    //  保存推荐列表
    List<Recommendation> scores = new ArrayList<>();

    // (2) 定义两个 map,用于保存每个商品的高分和低分的计数器,productId -> count
    Map<Integer, Integer> increMap = new HashMap<>();
    Map<Integer, Integer> decreMap = new HashMap<>();

    // (3) 遍历每个备选商品,计算和已评分商品的相似度
    for (Integer candidateProduct : candidateProducts) {
        for (Recommendation userRecentlyRating : userRecentlyRatings) {
            // (4) 从相似度矩阵中获取当前备选商品和当前已评分商品间的相似度
            double simScore = getProductsSimScore(candidateProduct, userRecentlyRating.getProductId(), simProducts);
            if (simScore > 0.4) {
                // (5) 按照公式进行加权计算,得到基础评分
                scores.add(new Recommendation(candidateProduct, simScore * userRecentlyRating.getScore()));
                if (userRecentlyRating.getScore() > 3) {
                    increMap.put(candidateProduct, increMap.getOrDefault(candidateProduct, 0) + 1);
                } else {
                    decreMap.put(candidateProduct, decreMap.getOrDefault(candidateProduct, 0) + 1);
                }
            }
        }
    }

    // (4) 根据公式计算所有的推荐优先级,首先以 productId 做 groupby
```

```java
        Map<Integer, List<Recommendation>> groupedScores = scores.stream().collect(Collec-
tors.groupingBy(Recommendation::getProductId));

        for (Map.Entry<Integer, List<Recommendation>> entry : groupedScores.entrySet()) {
            int productId = entry.getKey();
            List<Recommendation> scoreList = entry.getValue();
            double sum = 0.0;
            for (Recommendation score : scoreList) {
                sum += score.getScore();
            }
            double score = sum / scoreList.size() + log(increMap.getOrDefault(productId, 1))
- log(decreMap.getOrDefault(productId, 1));
            scores.add(new Recommendation(productId, score));
        }

        // (5) 返回推荐列表,按照得分排序
        List<Recommendation> result = new ArrayList<>(scores);
        return result.stream().sorted().collect(Collectors.toList());
    }

    // (4) 从相似度矩阵中获取当前备选商品和当前已评分商品间的相似度
    public static double getProductsSimScore(int product1, int product2, Map<Integer, Map<
Integer, Double>> simProducts) {
        if (simProducts.containsKey(product1)) {
            Map<Integer, Double> sims = simProducts.get(product1);
            if (sims.keySet().contains(product2)) {
                return sims.get(product2);
            } else {
                return 0.0;
            }
        } else {
            return 0.0;
        }
    }

    // log 函数计算
    public static double log(int m) {
        final int n = 10;
        return Math.log(m) / Math.log(n);
    }

    // 4.4.存储数据库到数据库
    public static void saveDataToMongoDB(int userId, List<Recommendation> streamRecs) {
        // (1) 定义存储数据库的链接
        MongoDatabase database = MongoDBUtil.getMongoDatabase("recommender");
        MongoCollection<Document> collection = database.getCollection("StreamRecs");
```

```java
        // (2) 按照 userId 查询并更新
        Document query = new Document("userId", userId);
        collection.deleteOne(query);
        List<Document> list = streamRecs.stream()
                .map(x -> new Document("userId", userId)
                        .append("recs", new Document("productId", x.getProductId())
                                .append("score", x.getScore())))
                .collect(Collectors.toList());
        // (3) 保存数据
        collection.insertMany(list);
    }

    // 创建 Jedis
    public static Jedis getJedis(String host, int port) {
        return  new Jedis(host, port);
    }
}
```

第13章

Flink 实现电商用户行为分析

电商平台的运营过程中，用户行为数据是极其重要的一环。这些数据能够反映出用户的购物习惯、需求以及偏好，对于电商平台来说，理解并分析这些数据是提升销售、优化产品、增强风险控制的关键。

本章内容综合运用 Flink 的各种 API，基于 EventTime 去处理基本的业务需求，并且灵活地使用底层的 processFunction，基于状态编程和 CEP 去处理更加复杂的情形。本项目的主要目标就是通过收集和分析大量的用户行为数据，获取有价值的商业指标。

13.1 电商用户行为实时分析系统概述

电商用户行为分析系统是一款基于 Apache Flink 流处理框架构建的实时分析系统。该系统针对电商业务特点，从统计分析、偏好统计和风险控制三个方面对用户行为进行深入分析，为电商企业提供有力的数据支持。

1. 统计分析

统计分析主要针对用户在电商平台的浏览、点击等行为进行统计。通过实时处理用户行为数据，系统能够实时展示热门商品、最近热门商品、分类热门商品以及流量统计等信息。此外，系统还支持对用户偏好进行统计，为推荐系统和用户画像提供数据支持。

2. 偏好统计

偏好统计主要关注用户在电商平台上的收藏、兴趣、评分、打标签等行为。通过分析用户的行为数据，系统能够挖掘用户的兴趣和需求，为电商企业提供个性化的推荐服务。此外，根据用户的行为数据，系统还能够构建用户画像，全面了解用户的需求和喜好，为精准营销提供支持。

3. 风险控制

风险控制主要针对电商业务中的欺诈行为进行监控。系统能够实时监控刷单、订单失效等恶意行为，保护企业的营销资金。同时，系统还能够监控恶意登录等异常行为，保障电商平台的安全稳定运行。通过风险控制模块，企业能够及时发现并处理潜在的欺诈行为，提高电商平台的运营效率。

总之，基于 Apache Flink 流处理框架构建的电商用户行为分析系统能够帮助电商企业全面了解用户需求、精准推荐商品以及防范业务风险。通过实时分析用户行为数据，企业能够提高运营效率、提升用户体验并保障业务安全。

13.2 电商用户行为分析系统架构

13.2.1 电商用户行为分析系统模块介绍

实时热门商品统计、实时流量统计、市场营销分析、恶意登录监控、订单支付实时监控……在现代商业运营中，实时数据分析已经成为企业成功的关键因素之一。本书将深入探讨如何实现实时热门商品统计、实时流量统计、市场营销分析、恶意登录监控以及订单支付实时监控等重要环节。通过学习本书的内容，你将能够更好地了解企业运营中的数据驱动决策，并提升企业的竞争力和盈利能力。

该项目主要涉及两个模块，实时统计分析模块和业务流程及风险控制模块，详细模块信息如下所示。

1. 实时统计分析模块

1）实时热门商品统计：这个模块用来实时统计电商平台上的热门商品。利用 Flink 的流处理能力，从商品数据流中提取出热门商品的信息，并进行统计。这个模块可以帮助电商企业了解用户最感兴趣的商品，从而优化产品展示和营销策略。

2）实时流量统计热门网页：这个模块用于实时统计电商平台上热门网页的流量。利用 Flink 的流处理能力，从网页数据流中提取出热门网页的信息，并进行统计。这个模块可以帮助电商企业了解用户最常访问的网页，从而优化网页设计和营销策略。

3）实时访问流量统计：这个模块用于实时统计电商平台上每秒的页面浏览量（PV）和用户浏览量（UV）。利用 Flink 的流处理能力，从用户行为数据流中提取出 PV 和 UV 的信息，并进行统计。这个模块可以帮助电商企业了解网站的流量情况，从而优化产品展示和营销策略。

4）市场营销分析市场推广 APP 统计：这个模块用于实时统计电商平台上 APP 在各个市场的推广情况。利用 Flink 的流处理能力，从 APP 数据流中提取出各个市场的推广信息，并进行统计。这个模块可以帮助电商企业了解 APP 在不同市场的推广效果，从而优化推广策略。

5）市场营销分析页面广告统计：这个模块用于实时统计电商平台上各个页面广告的点击情况。利用 Flink 的流处理能力，从广告数据流中提取出各个广告的点击信息，并进行统计。使用 CEP 技术来检测广告点击的异常行为并进行过滤。这个模块可以帮助电商企业了解广告的点击情况和优化广告策略。

2. 业务流程及风险控制模块

1）恶意登录监控：这个模块用于实时监控电商平台的恶意登录行为。从用户行为数据流中提取出恶意登录的信息，并及时进行干预和处理。这个模块可以帮助电商企业保护平台的安全性和稳定性，提高用户体验。

2）订单支付实时监控：这个模块用于实时监控电商平台的订单支付情况。从订单支付数据流中提取出异常支付的信息，并及时进行干预和处理。这个模块可以帮助电商企业保护资金流动的安全性，提高运营效率。

3）订单支付实时对账：这个模块用于实时对账电商平台的订单支付数据和银行系统支付数据的一致性。可以将订单支付数据流和银行支付数据流进行对账处理，并及时发现和处理不一致的情况。这个模块可以帮助电商企业提高资金流动的准确性，降低财务风险。

▶▶ 13.2.2　创建电商用户行为分析项目

新建一个电商用户行为分析项目，名为 jareny-bigdata-flink-user-behavior-analysis，该电商用户行为分析系统的内容主要包括实时统计分析模块和业务流程及风险控制模块两大部分，这两个模块的实现均使用实时的处理技术。

为了实现这些功能，电商用户分析系统添加了一些依赖包。其中，**Apache Flink** 是主要的大数据处理工具，用于实时处理和分析用户行为数据。该系统还使用了其他一些开源库和工具，在电商用户行为分析项目的 pom.xml 文件中添加依赖，代码如下。

```xml
<?xml version="1.0" encoding="UTF-8"?>
<project xmlns="http://maven.apache.org/POM/4.0.0"
    xmlns:xsi="http://www.w3.org/2001/XMLSchema-instance"
    xsi:schemaLocation="http://maven.apache.org/POM/4.0.0 http://maven.apache.org/xsd/maven-4.0.0.xsd">
    <modelVersion>4.0.0</modelVersion>

    <groupId>com.it.jareny.bigdata</groupId>
    <artifactId>jareny-bigdata-flink-user-behavior-analysis</artifactId>
    <version>1.0-SNAPSHOT</version>

    <properties>
        <project.build.sourceEncoding>UTF-8</project.build.sourceEncoding>
        <!--maven properties -->
        <maven.test.skip>false</maven.test.skip>
        <maven.javadoc.skip>false</maven.javadoc.skip>
        <!-- compiler settings properties -->
        <maven.compiler.source>1.8</maven.compiler.source>
        <maven.compiler.target>1.8</maven.compiler.target>
        <flink.version>1.13.3</flink.version>
        <commons-lang.version>2.5</commons-lang.version>
        <scala.binary.version>2.11</scala.binary.version>
        <spotless.version>2.4.2</spotless.version>
        <kafka.version>2.2.0</kafka.version>
    </properties>

    <dependencies>
        <!-- flink 公有依赖 -->
```

```xml
<dependency>
    <groupId>org.apache.flink</groupId>
    <artifactId>flink-java</artifactId>
    <version>${flink.version}</version>
</dependency>
<dependency>
    <groupId>org.apache.flink</groupId>
    <artifactId>flink-streaming-java_${scala.binary.version}</artifactId>
    <version>${flink.version}</version>
</dependency>
<dependency>
    <groupId>org.apache.flink</groupId>
    <artifactId>flink-clients_${scala.binary.version}</artifactId>
    <version>${flink.version}</version>
</dependency>
<dependency>
    <groupId>org.apache.flink</groupId>
    <artifactId>flink-json</artifactId>
    <version>${flink.version}</version>
</dependency>
<!-- flinkSql 依赖 -->
<dependency>
    <groupId>org.apache.flink</groupId>
    <artifactId>flink-table-api-java-bridge_${scala.binary.version}</artifactId>
    <version>${flink.version}</version>
</dependency>
<dependency>
    <groupId>org.apache.flink</groupId>
    <artifactId>flink-table-planner-blink_${scala.binary.version}</artifactId>
    <version>${flink.version}</version>
</dependency>
<!-- flink-kafka-connector -->
<dependency>
    <groupId>org.apache.kafka</groupId>
    <artifactId>kafka_${scala.binary.version}</artifactId>
    <version>${kafka.version}</version>
</dependency>
<dependency>
    <groupId>org.apache.flink</groupId>
    <artifactId>flink-connector-kafka_${scala.binary.version}</artifactId>
    <version>${flink.version}</version>
</dependency>
<!--lombok 插件依赖-->
<dependency>
    <groupId>org.projectlombok</groupId>
    <artifactId>lombok</artifactId>
```

第 13 章
Flink 实现电商用户行为分析

```xml
        <version>1.18.12</version>
    </dependency>
    <dependency>
        <groupId>commons-lang</groupId>
        <artifactId>commons-lang</artifactId>
        <version>${commons-lang.version}</version>
    </dependency>
    <!--Flink 默认使用的是 slf4j 记录日志,相当于一个日志的接口,这里使用 log4j 作为具体的日志实现-->
    <dependency>
        <groupId>org.slf4j</groupId>
        <artifactId>slf4j-log4j12</artifactId>
        <version>1.7.7</version>
        <scope>runtime</scope>
    </dependency>
    <dependency>
        <groupId>org.apache.logging.log4j</groupId>
        <artifactId>log4j-core</artifactId>
        <version>2.10.0</version>
    </dependency>
    <dependency>
        <groupId>log4j</groupId>
        <artifactId>log4j</artifactId>
        <version>1.2.17</version>
        <scope>runtime</scope>
    </dependency>
    <!-- 单元测试依赖 -->
    <dependency>
        <groupId>junit</groupId>
        <artifactId>junit</artifactId>
        <version>4.12</version>
        <!-- junit 的版本有 3.x, 4.x, 5.x。5.x 还没有发布, 现在都用 4.x -->
    </dependency>
    <!-- redis 依赖 -->
    <dependency>
        <groupId>redis.clients</groupId>
        <artifactId>jedis</artifactId>
        <version>3.3.0</version>
    </dependency>
    <!--flink-StarRocks-connector-->
    <dependency>
        <groupId>com.starrocks</groupId>
        <artifactId>flink-connector-starrocks</artifactId>
        <version>1.1.10_flink-1.13</version>
    </dependency>
    <!--flink-CEP-内库-->
```

```xml
    <dependency>
        <groupId>org.apache.flink</groupId>
        <artifactId>flink-cep_${scala.binary.version}</artifactId>
        <version>${flink.version}</version>
    </dependency>
    <!--mysql-cdc-->
    <dependency>
        <groupId>com.alibaba.ververica</groupId>
        <artifactId>flink-connector-mysql-cdc</artifactId>
        <version>1.2.0</version>
    </dependency>
    <dependency>
        <groupId>mysql</groupId>
        <artifactId>mysql-connector-java</artifactId>
        <version>5.1.47</version>
    </dependency>
</dependencies>
</project>
```

通过这些依赖包的支持，电商用户分析系统能够高效地处理大规模的数据，提供实时分析和报警功能，以及可视化分析和查询界面。该系统的用户可以轻松地了解用户行为、市场趋势和风险情况，从而做出更明智的商业决策。

13.3 实时热门商品统计

实时热门商品统计可以帮助企业了解当前最受欢迎的商品，从而更好地调整库存和销售策略。本书将介绍如何收集热门商品信息，如何根据商品热度排名进行统计，以及如何呈现给读者。此外，我们还将探讨如何利用数据分析和机器学习技术来预测未来商品趋势，以提前做好准备。

13.3.1 实时热门商品统计模块介绍

实时热门商品统计是一个基于 Apache Flink 的实时分析系统，用于统计电商平台上近 1 个小时内的热门商品，每 5 分钟更新一次，热门度使用浏览（PV）来衡量。该模块通过过滤所有用户行为数据中的浏览行为进行统计，并构建滑动窗口来计算每一种商品的访问数。然后根据滑动窗口的时间，统计出访问次数最多的 5 个商品。

13.3.2 实时热门商品统计模块实现

1. 项目目标

实时统计近 1 个小时内的热门商品，每 5 分钟更新一次。计算每一种商品的访问数，并根据访问次数排序。过滤出浏览行为进行统计，以准确衡量商品的热门度。使用滑动窗口实现数据的实时更新

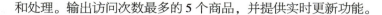

和处理。输出访问次数最多的 5 个商品，并提供实时更新功能。

2. 技术方案

使用 Apache Flink 作为流处理框架，实现数据的实时分析和处理。利用 Java 语言编写 Java 对象，用于存储和处理数据。利用 Flink 的流处理能力，对用户行为数据进行实时过滤和统计。构建滑动窗口，窗口长度为 1 小时，滑动距离为 5 分钟，计算每一种商品的访问数。使用 Flink 的并行处理能力，对数据进行分布式处理，提高计算效率。

3. 程序代码实现

1）创建用户行为对象。用户行为对象是指对用户在电商平台上的操作和活动进行建模和表示的对象。需要明确收集用户行为数据，例如用户的登录方式、浏览页面、点击行为、购买行为等。这些数据可以从 Web 服务器日志、业务系统日志等来源收集。

```java
package com.jareny.flink.userbehavior.analysis.entity;

import lombok.AllArgsConstructor;
import lombok.Data;
import lombok.NoArgsConstructor;

// 用户行为
@Data
@AllArgsConstructor
@NoArgsConstructor
public class UserBehavior {
    // 用户 ID
    private Long userId;
    // 商品 ID
    private Long itemId;
    // 商品分类 ID
    private Integer categoryId;
    // 用户行为（浏览,购买,加入购物车等）
    private String behavior;
    // 用户行为发生时间戳
    private Long timestamp;
}
```

2）创建统计输出对象是指将分析后的用户行为数据以统计学的形式呈现，以便用户能够快速了解和分析数据。根据业务需求和数据分析结果，确定需要输出的统计指标，例如平均值、总数、频率、百分比等。这些指标应该能够反映出用户行为的重要方面和趋势。

```java
package com.jareny.flink.userbehavior.analysis.entity;

import lombok.AllArgsConstructor;
import lombok.Data;
import lombok.NoArgsConstructor;
```

```java
// 热门品类输出
@Data
@AllArgsConstructor
@NoArgsConstructor
public class ItemViewCount {
    // 商品 ID
    private Long itemId;
    // 创建介绍时候
    private Long windowEnd;
    // 统计次数
    private Long count;
}
```

3) 计算热门商品实现。计算热门商品的程序主体代码包括以下步骤：收集商品销售数据、数据清洗和整理、计算热门商品、生成报告。代码如下。

```java
package com.jareny.flink.userbehavior.analysis.task;

import com.jareny.flink.userbehavior.analysis.entity.ItemViewCount;
import com.jareny.flink.userbehavior.analysis.entity.UserBehavior;
import org.apache.flink.api.common.functions.AggregateFunction;
import org.apache.flink.api.common.state.ListState;
import org.apache.flink.api.common.state.ListStateDescriptor;
import org.apache.flink.api.java.tuple.Tuple;
import org.apache.flink.configuration.Configuration;
import org.apache.flink.shaded.guava18.com.google.common.collect.Lists;
import org.apache.flink.streaming.api.datastream.DataStreamSource;
import org.apache.flink.streaming.api.datastream.SingleOutputStreamOperator;
import org.apache.flink.streaming.api.environment.StreamExecutionEnvironment;
import org.apache.flink.streaming.api.functions.KeyedProcessFunction;
import org.apache.flink.streaming.api.functions.timestamps.AscendingTimestampExtractor;
import org.apache.flink.streaming.api.functions.windowing.WindowFunction;
import org.apache.flink.streaming.api.windowing.time.Time;
import org.apache.flink.streaming.api.windowing.windows.TimeWindow;
import org.apache.flink.util.Collector;
import java.sql.Timestamp;
import java.util.ArrayList;
import java.util.Comparator;

/**
 * 1 小时内热门商品统计,每 5 分钟计算一次
 */
public class HotItemsTask {
    public static void main(String[] args) throws Exception {
        // 1.创建 Flink 运行环境
        StreamExecutionEnvironment env = StreamExecutionEnvironment.getExecutionEnvironment();
```

第13章 Flink 实现电商用户行为分析

```java
    // 设置 Flink 的并行度
    env.setParallelism(1);

    // 2.读取数据
    DataStreamSource<String> inputStream = env.readTextFile(
            "D:\\jareny\\bigdata\\jareny-bigdata-flink-user-behavior-analysis\\src\\main\\resources\\UserBehavior.csv");

    // 3.将数据转换为 UserBehavior 用户行为实体类对象,并设置事件时间
    SingleOutputStreamOperator<UserBehavior> dataStream = inputStream.map(line -> {
        String[] filed = line.split(",");
         return new UserBehavior(new Long(filed[0]), new Long(filed[1]), new Integer(filed[2]), filed[3], new Long(filed[4]));
     }).assignTimestampsAndWatermarks(new AscendingTimestampExtractor<UserBehavior>() {
        @Override
        public long extractAscendingTimestamp(UserBehavior userBehavior) {
            return userBehavior.getTimestamp() * 1000;
        }
    });

    // 4.筛选出 pv 的数据,按照商品 id 分组,划分滑动时间窗口,对每个窗口进行增量聚合,
        // 并将输出结果进行设定,指定格式 ItemViewCount
    SingleOutputStreamOperator<ItemViewCount> windowAggStream = dataStream
            .filter(data -> "pv".equals(data.getBehavior()))
            .keyBy("itemId")
            .timeWindow(Time.seconds(60), Time.seconds(30))
            // aggregate 第一个参数是窗口聚合的规则,第二个参数是定义输出的数据结构
            .aggregate(new CountAgg(), new WindowResultFunction());

    // 5.将同一个窗口的数据进行分组,最后设置定时输出
    SingleOutputStreamOperator<String> resultStream = windowAggStream
            .keyBy("windowEnd")
            .process(new TopNHotItems(5));
    resultStream.print();

    // 6.开始提交到 Flink 执行
    env.execute("hot items");
}

/**
 * 设定同一个商品数据的聚合方法
 * AggregateFunction<IN, ACC, OUT>
 * IN :输出类型
 * ACC:累加器类型
 * OUT:最后输出结果
```

```java
     */
    private static class CountAgg implements AggregateFunction<UserBehavior, Long, Long> {
        @Override
        public Long createAccumulator() {
            return 0L;
        }

        @Override
        public Long add(UserBehavior userBehavior, Long aLong) {
            return aLong + 1;
        }

        @Override
        public Long getResult(Long aLong) {
            return aLong;
        }

        @Override
        public Long merge(Long aLong, Long acc1) {
            return aLong + acc1;
        }
    }

    /**
     * 设定输出格式
     * WindowFunction<IN,OUT,KEY,W extends Window>
     * IN :输出类型,就是累加器最后输出类型
     * OUT:最后想要输出类型
     * KEY:Tuple 泛型,分组的 key,在这里是 itemId,窗口根据 itemId 聚合
     * W  :聚合的窗口,w.getEnd 就能拿到窗口的结束时间
     */
    private static class WindowResultFunction implements WindowFunction<Long, ItemViewCount, Tuple, TimeWindow> {
        @Override
        public void apply(Tuple tuple, TimeWindow timeWindow, Iterable<Long> iterable,
                          Collector<ItemViewCount> out) throws Exception {
            Long itemId = tuple.getField(0);
            Long end = timeWindow.getEnd();
            Long next = iterable.iterator().next();
            out.collect(new ItemViewCount(itemId, end, next));
        }
    }

    /**
     * 设置热门的 TOP 商品
     * KeyedProcessFunction<KEY,IN,OUT>
```

第 13 章
Flink 实现电商用户行为分析

```
 * KEY:分组 key 的类型
 * IN :输入的类型
 * OUT:输出的类型
 */
private static class TopNHotItems extends KeyedProcessFunction<Tuple, ItemViewCount, String> {
    private Integer topSize;

    public TopNHotItems(Integer topSize) {
        this.topSize = topSize;
    }

    ListState<ItemViewCount> itemViewCountListState;

    @Override
    public void open(Configuration parameters) throws Exception {
        itemViewCountListState = getRuntimeContext().getListState(new ListStateDescriptor<ItemViewCount>("Item-view-count-list", ItemViewCount.class));
    }

    @Override
    public void processElement(ItemViewCount count, KeyedProcessFunction<Tuple, ItemViewCount, String>.Context ctx, Collector<String> out) throws Exception {
        itemViewCountListState.add(count);
        // 注册一个定时器,在 1 毫秒之后运行,由于同一个窗口的结束时间是一样的,
        // 所以当时间变了,就说明同一个窗口的数据都添加进去了
        ctx.timerService().registerEventTimeTimer(count.getWindowEnd() + 1);
    }

    // 设定定时器任务
    @Override
    public void onTimer (long timestamp, KeyedProcessFunction<Tuple, ItemViewCount, String>.OnTimerContext ctx, Collector<String> out) throws Exception {
        // 将 ListState 转换成 ArrayList
        ArrayList<ItemViewCount> itemViewCounts = Lists.newArrayList(itemViewCountListState.get().iterator());

        // 排序
        itemViewCounts.sort(new Comparator<ItemViewCount>() {
            @Override
            public int compare(ItemViewCount o1, ItemViewCount o2) {
                return o2.getCount().intValue() - o1.getCount().intValue();
            }
        });

        // 将数据格式化
```

· 267 ·

```
            StringBuffer stringBuffer = new StringBuffer();
            stringBuffer.append("===========\n");
            stringBuffer.append("窗口结束时间:").append(new Timestamp(timestamp - 1)).append
("\n");

            for (int i = 0; i < Math.min(topSize, itemViewCounts.size()); i++) {
                ItemViewCount itemViewCount = itemViewCounts.get(i);
                stringBuffer.append("NO ").append(i + 1).append(":")
                        .append(" 商品 id = ").append(itemViewCount.getItemId())
                        .append(" 热门度 = ").append(itemViewCount.getCount())
                        .append("\n");
            }
            stringBuffer.append("===========\n\n");

            // 控制输出频率
            Thread.sleep(1000L);

            // 输出数据
            out.collect(stringBuffer.toString());
        }
    }
}
```

13.4 实时流量统计

电商分析系统中的实时热门页面统计功能,旨在帮助电商平台实时掌握用户最关注的页面和商品信息,以便更好地了解用户需求和行为,优化产品和服务。本书将介绍如何收集网站实时流量信息,如何分析流量趋势,以及如何根据流量做出相应的营销策略。

▶▶ 13.4.1 实时热门页面统计介绍

实时流量统计热门页面是一个基于 Apache Flink 的实时分析系统,用于从 Web 服务器的日志中统计实时的热门访问页面。每分钟将访问量最大的 5 个地址取出,每 5 秒更新一次。该模块将 apache 服务器日志中的时间转换为时间戳,作为 Event Time,并筛选出 GET 请求的网页,将请求资源的数据过滤掉。然后根据 URL 进行分组,构建滑动窗口,窗口长度为 1 分钟,滑动距离为 5 秒,进行增量聚合,并指定格式输出。最后根据窗口的时间分组,将同一个窗口的数据聚合,格式化输出。

▶▶ 13.4.2 实时热门页面统计模块实现

1. 项目目标

实时从 Web 服务器日志中统计热门访问页面。每分钟将访问量最大的 5 个地址取出,每 5 秒更新一次。将 apache 服务器日志中的时间转换为时间戳,作为 Event Time。筛选出 GET 请求的网页,将请

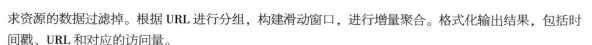

求资源的数据过滤掉。根据 URL 进行分组，构建滑动窗口，进行增量聚合。格式化输出结果，包括时间戳、URL 和对应的访问量。

2. 技术方案

使用 Apache Flink 作为流处理框架，实现数据的实时分析和处理。利用 Java 语言编写对象，用于存储和处理数据。利用 Flink 的流处理能力，对 Web 服务器日志进行实时解析和聚合。将 apache 服务器日志中的时间转换为时间戳，作为 Event Time，并利用该时间戳进行数据分组和聚合。筛选出 GET 请求的网页，将请求资源的数据过滤掉。根据 URL 进行分组，构建滑动窗口，窗口长度为 1 分钟，滑动距离为 5 秒，进行增量聚合。使用 Flink 的并行处理能力，对数据进行分布式处理，提高计算效率。

3. 程序代码实现

1）创建 Web 服务器日志模型。Web 服务器日志包含了用户与电商平台交互的详细信息，如页面浏览、点击行为、购买行为等。通过创建 Web 服务器日志模型，我们可以将这些信息提取出来，并进行处理和分析，以提取有价值的商业指标。

```java
package com.jareny.flink.userbehavior.analysis.entity;

import lombok.AllArgsConstructor;
import lombok.Data;
import lombok.NoArgsConstructor;

@Data
@AllArgsConstructor
@NoArgsConstructor
public class ApacheLogEvent {
    // 用户 IP
    private String ip;
    // 用户 ID
    private String userId;
    // 请求时间戳
    private Long timestamp;
    // 请求方法
    private String method;
    // 请求路径
    private String url;
}
```

2）创建登录统计模型。从电商平台的登录系统或用户数据库中收集登录相关的数据，包括用户 ID、登录时间、登录方式（如 PC 端、移动端）、登录来源（如直接输入网址、通过广告点击）等。

```java
package com.jareny.flink.userbehavior.analysis.entity;

import lombok.AllArgsConstructor;
import lombok.Data;
import lombok.NoArgsConstructor;
```

```java
// 登录事件
@Data
@AllArgsConstructor
@NoArgsConstructor
public class LoginEvent {
    // 用户 ID
    private Long userId;
    // 登录 IP
    private String ip;
    // 登录状态
    private String loginState;
    // 登录时间戳
    private Long timestamp;
}
```

3）热门网页统计代码实现。电商用户分析系统中的热门网页统计功能，可以帮助电商平台实时掌握用户最关注的网页信息，以便更好地了解用户需求和行为，优化产品和服务。实现热门网页统计功能时，有以下几个步骤：数据清洗和整理、数据提取和处理、数据存储和可视化。

```java
package com.jareny.flink.userbehavior.analysis.task;

import com.jareny.flink.userbehavior.analysis.entity.ApacheLogEvent;
import com.jareny.flink.userbehavior.analysis.entity.PageViewCount;
import org.apache.commons.compress.utils.Lists;
import org.apache.flink.api.common.functions.AggregateFunction;
import org.apache.flink.api.common.state.ListState;
import org.apache.flink.api.common.state.ListStateDescriptor;
import org.apache.flink.configuration.Configuration;
import org.apache.flink.streaming.api.datastream.DataStreamSource;
import org.apache.flink.streaming.api.datastream.SingleOutputStreamOperator;
import org.apache.flink.streaming.api.environment.StreamExecutionEnvironment;
import org.apache.flink.streaming.api.functions.KeyedProcessFunction;
import org.apache.flink.streaming.api.functions.timestamps.BoundedOutOfOrdernessTimestampExtractor;
import org.apache.flink.streaming.api.functions.windowing.WindowFunction;
import org.apache.flink.streaming.api.windowing.time.Time;
import org.apache.flink.streaming.api.windowing.windows.TimeWindow;
import org.apache.flink.util.Collector;

import java.sql.Timestamp;
import java.text.SimpleDateFormat;
import java.util.ArrayList;
import java.util.regex.Pattern;

/**
```

第 13 章 Flink 实现电商用户行为分析

```java
 * 热门网页统计
 */
public class HotPagesTask {
    public static void main(String[] args) throws Exception {
        // 1.创建 Flink 运行环境
        StreamExecutionEnvironment env = StreamExecutionEnvironment.getExecutionEnvironment();
        // 设置 Flink 的并行度
        env.setParallelism(1);

        // 2.读取数据
        DataStreamSource<String> inputStream = env.readTextFile(
                "D:\\jareny\\bigdata\\jareny-bigdata-flink-user-behavior-analysis\\src\\main\\resources\\apache.log");

        // 3.将字符串格式的 DataSream 转换为登录日志格式,并设置 EventTime 和 watermark
        SingleOutputStreamOperator<ApacheLogEvent> dataStream = inputStream.map(line -> {
            String[] fields = line.split(" ");
            SimpleDateFormat simpleDateFormat = new SimpleDateFormat("dd/MM/yyyy:HH:mm:ss");
            long time = simpleDateFormat.parse(fields[3]).getTime();
            return new ApacheLogEvent(fields[0], fields[1], time, fields[5], fields[6]);
        }).assignTimestampsAndWatermarks(
                new BoundedOutOfOrdernessTimestampExtractor<ApacheLogEvent>(Time.seconds(1)) {
                    @Override
                    public long extractTimestamp(ApacheLogEvent element) {
                        return element.getTimestamp();
                    }
                });

        /**
         * 4.筛选数据,将请求和请求网页的数据筛选出来
         * 之后根据 url 分组,设置滑动窗口(窗口大小为 1 分钟,滑动距离为 5 秒)
         * 最后增量聚合,设置指定格式输出
         */
        SingleOutputStreamOperator<PageViewCount> resStream = dataStream
                .filter(data -> "GET".equals(data.getMethod()))
                .filter(data -> {
                    String regex = "^((?!\\\\.(css|js|png|ico)$).)*$";
                    return Pattern.matches(regex, data.getUrl());
                })
                .keyBy(ApacheLogEvent::getUrl)
                .timeWindow(Time.minutes(1), Time.seconds(5))
                // 统计输出和窗口函数
                .aggregate(new PageCountAgg(), new PageView());

        /**
         * 5.根据窗口的最后时间进行分组,对分组之后的数据格式化输出
```

```java
     */
    SingleOutputStreamOperator<String> resultStream = resStream
            .keyBy(PageViewCount::getWindowEnd)
            .process(new MyProcessFunc());

    resultStream.print();

    // 6.开始提交到 Flink 执行
    env.execute();
}

/**
 * 统计方式
 */
public static class PageCountAgg implements AggregateFunction<ApacheLogEvent, Long, Long>{

    @Override
    public Long createAccumulator() {
        return 0L;
    }

    @Override
    public Long add(ApacheLogEvent apacheLogEvent, Long aLong) {
        return aLong+1;
    }

    @Override
    public Long getResult(Long aLong) {
        return aLong;
    }

    @Override
    public Long merge(Long aLong, Long acc1) {
        return aLong+acc1;
    }
}

/**
 * 窗口函数
 */
public static class PageView implements WindowFunction<Long, PageViewCount,String, TimeWindow>{
    @Override
    public void apply(String s, TimeWindow window, Iterable<Long> input, Collector<PageViewCount> out)   {
        out.collect(new PageViewCount(s, window.getEnd(), input.iterator().next()));
```

        }
    }

    /**
     * 数据格式化输出
     */
    public static class MyProcessFunc extends KeyedProcessFunction<Long, PageViewCount, String> {
        private ListState<PageViewCount> list;

        @Override
        public void onTimer(long timestamp, KeyedProcessFunction<Long, PageViewCount,
                String>.OnTimerContext ctx, Collector<String> out) throws Exception {
            ArrayList<PageViewCount> pageViewCountArrayList = Lists.newArrayList(list.get().iterator());

            pageViewCountArrayList.sort((p1,p2) -> p2.getCount().intValue() - p1.getCount().intValue());

            StringBuffer result = new StringBuffer();
            result.append("====================\n");
            result.append("窗口结束时间").append(new Timestamp(timestamp - 1)).append("\n");

            for (int i = 0; i < Math.min(5, pageViewCountArrayList.size()); i++) {
                PageViewCount pageView = pageViewCountArrayList.get(i);
                result.append(pageView.getUrl()).append("   ").append(pageView.getCount()).append("\n");
            }
            result.append("===============\n\n\n");

            Thread.sleep(1000);

            out.collect(result.toString());
        }

        @Override
        public void open(Configuration parameters) throws Exception {
            list = getRuntimeContext().getListState(
                    new ListStateDescriptor<PageViewCount>("PageViewCount",PageViewCount.class));
        }

        @Override
        public void processElement(PageViewCount value, KeyedProcessFunction<Long, PageViewCount, String>.Context ctx,
                                   Collector<String> out) throws Exception {
            list.add(value);
```
```

```
            ctx.timerService().registerEventTimeTimer(value.getWindowEnd()+1);
        }
    }
}
```

13.4.3 实时每小时访问量统计模块介绍

实时流量统计可以展示每小时访问量，帮助企业实时了解网站或应用程序的流量情况。通过分析每小时的访问量数据，企业可以更好地了解用户行为和需求，并制定更有效的营销策略。

实时流量统计：PV 和 UV 是一个基于 Apache Flink 的实时分析系统，用于从埋点日志中统计实时的 PV（页面浏览量）和 UV（用户浏览量）。每小时将访问量进行累计，并对用户进行去重。该模块通过直接对数据筛选过滤之后，对 PV 行为进行 sum 累计，而对于 UV 行为，利用 Set 数据结构进行去重。针对超大规模的数据，考虑使用布隆过滤器进行去重。

13.4.4 实时每小时访问量统计模块实现

1. 项目目标

实时从埋点日志中统计 PV 和 UV。每小时更新一次访问量累计，并对用户进行去重。对 PV 行为进行 sum 累计，对 UV 行为进行去重处理。针对超大规模的数据，使用布隆过滤器进行去重优化。输出实时 PV 和 UV 数据，支持数据可视化展示。

2. 技术方案

使用 Apache Flink 作为流处理框架，实现数据的实时分析和处理。利用 Java 语言编写对象，用于存储和处理数据。利用 Flink 的流处理能力，对埋点日志进行实时解析和聚合。对 PV 行为进行 sum 累计，通过直接筛选过滤进行累计。利用 Set 数据结构进行 UV 行为的去重处理。对于超大规模的数据，使用布隆过滤器进行去重优化。使用 Flink 的并行处理能力，对数据进行分布式处理，提高计算效率。

3. 程序代码实现

电商用户分析系统中的实时每小时访问量统计模块，旨在提供实时的每小时访问量数据，以便平台能够及时掌握用户访问情况，并做出相应的策略调整。该模块的代码实现，包括收集数据、数据清洗和整理、数据统计、数据存储、数据可视化等步骤。代码如下：

```
package com.jareny.flink.userbehavior.analysis.task;

import com.jareny.flink.userbehavior.analysis.entity.UserBehavior;
import org.apache.flink.api.common.functions.MapFunction;
import org.apache.flink.api.java.tuple.Tuple2;
import org.apache.flink.streaming.api.datastream.DataStreamSource;
import org.apache.flink.streaming.api.datastream.SingleOutputStreamOperator;
```

```java
import org.apache.flink.streaming.api.environment.StreamExecutionEnvironment;
import org.apache.flink.streaming.api.functions.timestamps.AscendingTimestampExtractor;
import org.apache.flink.streaming.api.windowing.time.Time;

/**
 * 统计每小时的访问量
 */
public class PageViewTask {
    public static void main(String[] args) throws Exception {
        // 1.创建 Flink 运行环境
        StreamExecutionEnvironment env = StreamExecutionEnvironment.getExecutionEnvironment();
        env.setParallelism(1);

        // 2.读取数据
        DataStreamSource<String> inputSream = env.readTextFile(
                "D:\\jareny\\bigdata\\jareny-bigdata-flink-user-behavior-analysis\\src\\main\\resources\\UserBehavior.csv");

        // 3.将数据转换为 UserBehavior 用户行为实体类对象,并设置事件时间
        SingleOutputStreamOperator<UserBehavior> dataStream = inputSream.map(line -> {
            String[] field = line.split(",");
            return new UserBehavior(new Long(field[0]), new Long(field[1]), new Integer(field[2]), field[3], new Long(field[4]));
        }).assignTimestampsAndWatermarks(new AscendingTimestampExtractor<UserBehavior>() {
            @Override
            public long extractAscendingTimestamp(UserBehavior element) {
                return element.getTimestamp() * 1000L;
            }
        });

        // 4.根据用户的行为统计页面访问量
        SingleOutputStreamOperator<Tuple2<String, Long>> result = dataStream
                .filter(data -> "pv".equals(data.getBehavior()))
                .map(new MapFunction<UserBehavior, Tuple2<String, Long>>() {
                    @Override
                    public Tuple2<String, Long> map(UserBehavior userBehavior) throws Exception {
                        return new Tuple2<>("pv", 1L);
                    }
                })
                .keyBy(data -> data.f0)
                .timeWindow(Time.hours(1))
                .sum(1);

        result.print();

        // 5.开始提交到 Flink 执行
```

```
        env.execute();
    }
}
```

13.4.5 实时用户访问量统计模块介绍

实时流量统计可以提供用户访问量的实时数据。通过这些数据，企业可以了解用户对网站或应用程序的访问情况和需求，以便制定更有效的营销策略。

企业还可以通过在应用程序或网站中嵌入代码来实现实时流量统计。通过在关键位置添加跟踪代码，企业可以实时收集用户行为数据，包括页面浏览量、停留时间、跳出率等。这些数据可以帮助企业更好地了解用户需求和行为，及时调整网站或应用程序的设计和功能。

总之，实时流量统计可以帮助企业更好地了解用户需求和行为，制定更有效的营销策略。通过使用统计工具或嵌入代码等方式，企业可以实时收集和分析用户访问量等关键数据，为运营决策提供有力支持。

13.4.6 实时用户访问量统计模块实现

1. 项目目标

实时收集用户访问数据，确保数据的实时性和准确性。对收集到的数据进行清洗、去重和格式转换，得到格式化的数据。实现实时用户访问量统计功能，包括每小时访问量、用户数等指标的计算和输出。提供数据可视化展示功能，将统计结果以图表的形式展示出来，以便用户更直观地了解数据。确保系统的稳定性和安全性，满足高并发和大数据量的处理需求。

2. 技术方案

可以选择使用日志分析或 API 接口等方式来收集用户访问数据。日志分析可以获取更全面的用户访问数据，但需要处理大量的日志数据；API 接口可以获取更实时的数据，但需要开发相应的接口程序。可以使用 Java 编程语言和 Flink 对数据进行清洗、去重、格式转换等操作。同时，也可以使用 ETL 工具来进行自动化数据处理。监控系统或告警系统用来接收统计结果，并根据设定的阈值及时触发告警。同时，也可以将统计结果输出到可视化大屏等展示系统中，以便更直观地了解访问情况。

3. 程序代码实现

电商用户行为分析中的用户访问量实时统计功能，可以帮助平台实时掌握用户访问情况，以便及时发现异常情况并做出相应的策略调整。下面是该功能的代码实现步骤，包括数据收集、数据预处理、数据存储、数据统计、数据输出可视化展示。代码如下。

```
package com.jareny.flink.userbehavior.analysis.task;

import com.jareny.flink.userbehavior.analysis.entity.PageViewCount;
import com.jareny.flink.userbehavior.analysis.entity.UserBehavior;
import org.apache.flink.streaming.api.TimeCharacteristic;
import org.apache.flink.streaming.api.datastream.DataStreamSource;
```

```java
import org.apache.flink.streaming.api.datastream.SingleOutputStreamOperator;
import org.apache.flink.streaming.api.environment.StreamExecutionEnvironment;
import org.apache.flink.streaming.api.functions.timestamps.AscendingTimestampExtractor;
import org.apache.flink.streaming.api.functions.windowing.AllWindowFunction;
import org.apache.flink.streaming.api.windowing.time.Time;
import org.apache.flink.streaming.api.windowing.windows.TimeWindow;
import org.apache.flink.util.Collector;
import java.util.HashSet;

/**
 * 统计用户访问量
 */
public class UserViewTask {
    public static void main(String[] args) throws Exception {
        // 1.创建Flink运行环境
        StreamExecutionEnvironment env = StreamExecutionEnvironment.getExecutionEnvironment();
        env.setParallelism(1);

        // 2.读取数据
        DataStreamSource<String> inputStream = env.readTextFile(
                "D:\\jareny\\bigdata\\jareny-bigdata-flink-user-behavior-analysis\\src\\main\\resources\\UserBehavior.csv");

        // 3.将数据转换为UserBehavior用户行为实体类对象,并设置事件时间
        SingleOutputStreamOperator<UserBehavior> dataStream = inputStream.map(line -> {
            String[] fields = line.split(",");
            return new UserBehavior(new Long(fields[0]), new Long(fields[1]), new Integer(fields[2]), fields[3], new Long(fields[4]));
        }).assignTimestampsAndWatermarks(new AscendingTimestampExtractor<UserBehavior>() {
            @Override
            public long extractAscendingTimestamp(UserBehavior element) {
                return element.getTimestamp() * 1000L;
            }
        });

        // 4.实现一个自定义全窗口函数,用于统计输出
        SingleOutputStreamOperator<PageViewCount> result = dataStream
                .filter(data -> "pv".equals(data.getBehavior()))
                .timeWindowAll(Time.hours(1))
                .apply(new UvCountResult());

        result.print();

        // 5.开始提交到Flink执行
        env.execute();
```

```java
    }
    // 实现一个自定义全窗口函数
    public static class UvCountResult implements AllWindowFunction<UserBehavior, PageViewCount, TimeWindow>{
        @Override
        public void apply(TimeWindow window, Iterable<UserBehavior> values, Collector<PageViewCount> out) throws Exception {
            // 定义一个 Set 结构,保存窗口中所有的 userid,自动去重
            HashSet<Long> uidSet = new HashSet<>();
            for (UserBehavior value : values) {
                uidSet.add(value.getUserId());
            }
            out.collect(new PageViewCount("uv",window.getEnd(),(long) uidSet.size()));
        }
    }
}
```

13.5 市场营销分析

13.5.1 市场营销分析——市场推广统计模块介绍

市场营销分析能够帮助企业了解市场趋势和竞争对手情况,制定更具针对性的营销策略。通过数据挖掘和可视化技术,企业可以深入洞察客户需求和市场变化,及时调整营销策略,提高市场竞争力。

13.5.2 市场营销分析——市场推广统计模块实现

市场营销分析——APP 市场推广统计是一个基于 Apache Flink 的实时分析系统,用于从埋点日志中统计 APP 市场推广的数据指标,并按照不同的推广渠道进行数据统计。该模块通过过滤日志中的用户行为,按照不同的渠道进行数据统计,并利用 Process Function 进行处理,得到自定义的输出数据信息。

1. 代码实现思路

从埋点日志中实时统计 APP 市场推广的数据指标。按照不同的推广渠道分别进行数据统计。输出自定义的输出数据信息,包括时间戳、推广渠道和相应的数据指标。支持数据可视化展示,帮助企业了解 APP 在不同市场的推广效果。

2. 技术方案

使用 Apache Flink 作为流处理框架,实现数据的实时分析和处理。利用 Java 语言编写对象,用于

存储和处理数据。利用 Flink 的流处理能力，对埋点日志进行实时解析和聚合。使用 Process Function 进行数据处理，得到自定义的输出数据信息。使用 Flink 的并行处理能力，对数据进行分布式处理，提高计算效率。

3. 代码实现

1）创建市场分析用户行为对象。市场分析用户行为对象是指将用户行为数据与市场数据相结合，以了解用户和市场的关系。

```java
package com.jareny.flink.userbehavior.analysis.entity;

import lombok.AllArgsConstructor;
import lombok.Data;
import lombok.NoArgsConstructor;

@Data
@AllArgsConstructor
@NoArgsConstructor
public class MarketingUserBehavior {
    // 用户 Id
    private Long userId;
    // 用户行为
    private String behavior;
    // 渠道
    private String channel;
    // 时间戳
    private Long timestamp;
}
```

2）创建渠道统计模型。渠道统计模型是指对用户从不同渠道进入电商平台的流量进行统计和分析，以了解不同渠道的效果和贡献。

```java
package com.jareny.flink.userbehavior.analysis.entity;

import lombok.AllArgsConstructor;
import lombok.Data;
import lombok.NoArgsConstructor;

@Data
@AllArgsConstructor
@NoArgsConstructor
public class ChannelPromotionCount {
    // 渠道
    private String channel;
    // 用户行为
    private String behavior;
    // 窗口结束时间
    private String windowEnd;
```

```java
    // 统计次数
    private Long count;
}
```

3) 市场营销代码实现。市场营销是电商用户行为分析的一个重要应用领域，通过精准的市场营销策略，可以更好地吸引用户、提高转化率、增加销售额。下面是一段描述电商用户行为分析的市场营销代码实现。

```java
package com.jareny.flink.userbehavior.analysis.task;

import com.jareny.flink.userbehavior.analysis.entity.ChannelPromotionCount;
import com.jareny.flink.userbehavior.analysis.entity.MarketingUserBehavior;
import org.apache.flink.api.common.functions.AggregateFunction;
import org.apache.flink.api.java.functions.KeySelector;
import org.apache.flink.api.java.tuple.Tuple2;;
import org.apache.flink.streaming.api.datastream.SingleOutputStreamOperator;
import org.apache.flink.streaming.api.environment.StreamExecutionEnvironment;
import org.apache.flink.streaming.api.functions.source.SourceFunction;
import org.apache.flink.streaming.api.functions.timestamps.AscendingTimestampExtractor;
import org.apache.flink.streaming.api.functions.windowing.WindowFunction;
import org.apache.flink.streaming.api.windowing.time.Time;
import org.apache.flink.streaming.api.windowing.windows.TimeWindow;
import org.apache.flink.util.Collector;
import java.util.Arrays;
import java.util.List;
import java.util.Random;

/**
 * APP 市场推广统计,不分渠道
 */
public class AppMarketingTask {
    public static void main(String[] args) throws Exception {
        //1.创建 Flink 运行环境
        StreamExecutionEnvironment env = StreamExecutionEnvironment.getExecutionEnvironment();
        env.setParallelism(1);

        //2.读取数据,使用自定义数据源
        SingleOutputStreamOperator<MarketingUserBehavior> dataStream = env.addSource(new MySource())
                .assignTimestampsAndWatermarks(new AscendingTimestampExtractor<MarketingUserBehavior>() {
                    @Override
                    public long extractAscendingTimestamp(MarketingUserBehavior element) {
                        return element.getTimestamp();
                    }
                });
```

第13章 Flink 实现电商用户行为分析

```java
        // 3.根据用户行为分组
        SingleOutputStreamOperator<ChannelPromotionCount> result = dataStream
                .filter(data -> !"UNINSTALL".equals(data.getBehavior()))
                .keyBy(new KeySelector<MarketingUserBehavior, Tuple2<String, String>>() {
                    @Override
                    public Tuple2<String, String> getKey(MarketingUserBehavior value) throws Exception {
                        return new Tuple2<>("total", "total");
                    }
                })
                .timeWindow(Time.seconds(30), Time.seconds(10))
                .aggregate(new MyAggFun(), new WindowFun());

        result.print();
        // 4.开始提交到 Flink 执行
        env.execute();
    }

    // AggregateFunction 它会自动对相同的 key 进行聚合,不需要再处理 key
    public static class MyAggFun implements AggregateFunction<MarketingUserBehavior, Long, Long>{

        @Override
        public Long createAccumulator() {
            return 0L;
        }

        @Override
        public Long add(MarketingUserBehavior marketingUserBehavior, Long aLong) {
            return aLong+1;
        }

        @Override
        public Long getResult(Long aLong) {
            return aLong;
        }

        @Override
        public Long merge(Long aLong, Long acc1) {
            return acc1+aLong;
        }
    }

    public static class WindowFun implements WindowFunction<Long,ChannelPromotionCount,Tuple2<String,String>,TimeWindow>{
        @Override
```

```java
    public void apply(Tuple2<String, String> tuple2, TimeWindow window,
                      Iterable<Long> input, Collector<ChannelPromotionCount> out) throws Exception {
        out.collect(new ChannelPromotionCount(tuple2.f0,tuple2.f1,String.valueOf(window.getEnd()),input.iterator().next()));
    }
}

// 设定自定义数据源,实现 SourceFunction
public static class MySource implements SourceFunction<MarketingUserBehavior> {
    Boolean running = true;

    // 定义用户行为和渠道的范围
    List<String> behaviorList = Arrays.asList("CLICK", "DOWNLOAD", "INSTALL", "UNINSTALL");
    List<String> channelList = Arrays.asList("app store", "wechat", "weibo");

    Random  random= new Random();

    @Override
    public void run(SourceContext<MarketingUserBehavior> ctx) throws Exception {
        while (running){
            // 随机生成所有字段
            long id = random.nextLong();
            String behavior = behaviorList.get(random.nextInt(behaviorList.size()));
            String channel = channelList.get(random.nextInt(channelList.size()));
            long timestamp = System.currentTimeMillis();

            // 发出数据
            ctx.collect(new MarketingUserBehavior(id,behavior,channel,timestamp));

            Thread.sleep(100L);
        }
    }

    @Override
    public void cancel() {
        running = false;
    }
}
```

13.5.3 市场营销分析——市场页面广告统计模块介绍

市场营销分析中的页面广告统计可以帮助企业了解页面广告的展示和点击情况,以便优化广告投放和营销策略。

13.5.4 市场营销分析——市场页面广告模块实现

市场营销分析——页面广告统计是一个基于 Apache Flink 的实时分析系统，用于从埋点日志中统计每小时页面广告的点击量，并按照不同省份进行划分。该模块根据省份进行分组，创建长度为 1 小时、滑动距离为 5 秒的时间窗口进行统计，并利用 Process Function 进行黑名单过滤，检测用户对同一广告的点击量，如果超过上限，则将用户信息以侧输出流输出到黑名单中。每 5 秒刷新一次统计数据，并支持数据可视化展示。

1. 项目目标

从埋点日志中实时统计每小时页面广告的点击量，每 5 秒更新一次。按照省份对广告点击量进行划分和统计。过滤刷单式的频繁点击行为，并将该用户加入黑名单。输出实时广告点击量数据，包括时间戳、省份和相应的广告点击量。支持数据可视化展示，帮助企业了解不同省份的广告点击情况。

2. 技术方案

使用 Apache Flink 作为流处理框架，实现数据的实时分析和处理。利用 Java 语言编写 POJO（Plain Old Java Object，普通老式 Java 对象），用于存储和处理数据。利用 Flink 的流处理能力，对埋点日志进行实时解析和聚合。根据省份进行分组，创建长度为 1 小时、滑动距离为 5 秒的时间窗口进行统计。使用 Process Function 进行数据处理，检测用户对同一广告的点击量，如果超过上限，则将用户信息以侧输出流输出到黑名单中。使用 Flink 的并行处理能力，对数据进行分布式处理，提高计算效率。

3. 代码实现

1）创建广告点击事件对象。广告点击事件对象是通过对用户在电商平台上点击广告的行为进行追踪和分析，了解广告效果、用户兴趣和行为模式的重要手段。

```java
package com.jareny.flink.userbehavior.analysis.entity;

import lombok.AllArgsConstructor;
import lombok.Data;
import lombok.NoArgsConstructor;

// 每小时页面广告的点击量
@Data
@AllArgsConstructor
@NoArgsConstructor
public class AdClickEvent {
    // 用户 ID
    private Long userId;
    // 广告 ID
    private Long adId;
    // 用户行为
```

```java
    private String province;
    // 用户所在城市
    private String city;
    // 时间戳
    private Long timestamp;
}
```

2）创建黑名单用户对象。黑名单用户对象是指那些被电商平台列为不良用户或恶意用户的名单。这些用户通常因为违反平台规定、有欺诈行为或其他不良行为而被列入黑名单。它主要用于管理和识别不良用户，以保护企业营销资金、提高活动运营效果和保障用户体验。

```java
package com.jareny.flink.userbehavior.analysis.entity;

import lombok.AllArgsConstructor;
import lombok.Data;
import lombok.NoArgsConstructor;

// 用户黑名单
@Data
@AllArgsConstructor
@NoArgsConstructor
public class BlackListUserWarning {
    // 用户 ID
    private Long userId;
    // 广告 ID
    private Long adId;
    // 警告行为
    private String warningMsg;
}
```

3）广告点击事件统计代码。电商用户行为分析广告点击事件统计是一段用于统计和分析电商用户点击广告事件的程序代码。通过该代码，可以追踪和记录用户点击广告的相关信息，以便了解用户对广告的兴趣和反应，以及优化广告投放策略。代码如下。

```java
package com.jareny.flink.userbehavior.analysis.task;

import com.jareny.flink.userbehavior.analysis.entity.AdClickEvent;
import com.jareny.flink.userbehavior.analysis.entity.AdCountViewByProvince;
import org.apache.flink.api.common.functions.AggregateFunction;
import org.apache.flink.streaming.api.datastream.DataStreamSource;
import org.apache.flink.streaming.api.datastream.SingleOutputStreamOperator;
import org.apache.flink.streaming.api.environment.StreamExecutionEnvironment;
import org.apache.flink.streaming.api.functions.timestamps.BoundedOutOfOrdernessTimestampExtractor;
import org.apache.flink.streaming.api.functions.windowing.WindowFunction;
import org.apache.flink.streaming.api.windowing.time.Time;
import org.apache.flink.streaming.api.windowing.windows.TimeWindow;
```

第 13 章
Flink 实现电商用户行为分析

```java
import org.apache.flink.util.Collector;

/**
 * 广告点击次数统计
 */
public class AdStatisticsByProvinceTask {
    public static void main(String[] args) throws Exception {
        // 1.创建 Flink 运行环境
        StreamExecutionEnvironment env = StreamExecutionEnvironment.getExecutionEnvironment();
        env.setParallelism(1);

        // 2.读取数据
        DataStreamSource<String> inputStream = env.readTextFile("D:\\IdeaProjects\\UserBehaviorAnalysis\\AdClickEvent\\AdClickLog.csv");

        // 3.将数据转换为 AdClickEvent 用户点击广告行为实体类对象,并设置事件时间
        SingleOutputStreamOperator<AdClickEvent> dataStream = inputStream.map(line -> {
            String[] field = line.split(",");
            return new AdClickEvent(new Long(field[0]), new Long(field[1]), field[2], field[3], new Long(field[4]));
        }).assignTimestampsAndWatermarks(new BoundedOutOfOrdernessTimestampExtractor<AdClickEvent>(Time.milliseconds(200)) {
            @Override
            public long extractTimestamp(AdClickEvent element) {
                return element.getTimestamp() * 1000L;
            }
        });

        // 4.统计每个省点击广告的次数,滑动窗口
        SingleOutputStreamOperator<AdCountViewByProvince> result = dataStream
                .keyBy(AdClickEvent::getProvince)
                .timeWindow(Time.minutes(10), Time.minutes(2))
                .aggregate(new MyAgg(), new WindowFun());
        result.print();

        // 5.开始提交到 Flink 执行
        env.execute();

    }

    public static class MyAgg implements AggregateFunction<AdClickEvent,Long,Long>{
        @Override
        public Long createAccumulator() {
            return 0L;
        }
```

```java
    @Override
    public Long add(AdClickEvent adClickEvent, Long aLong) {
        return aLong+1;
    }

    @Override
    public Long getResult(Long aLong) {
        return aLong;
    }

    @Override
    public Long merge(Long aLong, Long acc1) {
        return acc1+aLong;
    }
}

    public static class WindowFun implements WindowFunction<Long, AdCountViewByProvince, String, TimeWindow>{
        @Override
         public void apply(String s, TimeWindow window, Iterable<Long> input, Collector<AdCountViewByProvince> out) throws Exception {
                out.collect(new AdCountViewByProvince(s,String.valueOf(window.getEnd()), input.iterator().next()));
        }
    }
}
```

13.6 恶意登录监控

13.6.1 恶意登录监控模块描述

用户登录失败监控与异常检测是一个基于 Apache Flink 的实时分析系统，用于检测用户登录失败行为，并判断是否存在恶意攻击的可能性。该模块将用户的登录失败行为存入 ListState，并设定定时器以两秒为间隔触发，检查 ListState 中记录的失败登录次数。同时，为提高检测精度，考虑使用 CEP 库实现事件流的模式匹配。

13.6.2 恶意登录监控模块实现

1. 项目目标

实时监控用户登录失败行为，并记录相关信息。在两秒内连续登录失败的情况下，触发报警机制。通过模式匹配，精确检测出具有恶意攻击可能的用户行为。输出报警信息，提醒管理员进行干预和处理。提高系统安全性，有效防止恶意攻击行为。

2. 技术方案

使用 Apache Flink 作为流处理框架，实现数据的实时分析和处理。利用 Java 语言编写对象，用于存储和处理数据。利用 Flink 的 ListState 功能，将用户登录失败行为进行状态存储。设定定时器以两秒为间隔触发，检查 ListState 中记录的失败登录次数。使用 CEP 库实现事件流的模式匹配，提高检测精度。使用 Flink 的并行处理能力，对数据进行分布式处理，提高计算效率。

3. 代码实现

1）创建登录事件对象。电商用户行为分析中的登录事件对象是指用户在电商平台上进行登录操作的相关信息。通过该对象，可以了解用户登录的次数、时间、位置、设备等信息，以进一步分析用户的身份特征、行为模式和购买习惯。

```java
package com.jareny.flink.userbehavior.analysis.entity;

import lombok.AllArgsConstructor;
import lombok.Data;
import lombok.NoArgsConstructor;

// 登录事件
@Data
@AllArgsConstructor
@NoArgsConstructor
public class LoginEvent {
    // 用户 ID
    private Long userId;
    // 登录 IP
    private String ip;
    // 登录状态
    private String loginState;
    // 登录时间戳
    private Long timestamp;
}
```

2）创建登录告警对象。电商用户行为分析中的登录告警对象是指针对用户登录行为进行监控和预警的相关信息。通过该对象，可以及时发现异常登录行为，如异地登录、多次尝试登录等，以便采取相应的措施进行防范和处理。

```java
package com.jareny.flink.userbehavior.analysis.entity;

import lombok.AllArgsConstructor;
import lombok.Data;
import lombok.NoArgsConstructor;

// 登录失败告警
@Data
```

```java
@AllArgsConstructor
@NoArgsConstructor
public class LoginFailWarning {
    // 用户 ID
    private Long userId;
    // 第一次登录失败的时间
    private Long firstFailTime;
    // 最后一次登录失败的时间
    private Long lastFailTime;
    // 告警信息
    private String warningMsg;
}
```

3) 恶意登记监控代码实现。电商用户行为分析中的恶意登记监控代码实现是指通过编程技术对恶意用户在电商平台上恶意注册账号的行为进行监控和防范。通过该代码实现，可以有效地防止恶意用户在电商平台上进行恶意注册，保护企业的营销资金和正常用户的权益。

```java
package com.jareny.flink.userbehavior.analysis.task;

import com.jareny.flink.userbehavior.analysis.entity.LoginEvent;
import com.jareny.flink.userbehavior.analysis.entity.LoginFailWarning;
import org.apache.flink.api.common.state.ListState;
import org.apache.flink.api.common.state.ListStateDescriptor;
import org.apache.flink.api.common.state.ValueState;
import org.apache.flink.api.common.state.ValueStateDescriptor;
import org.apache.flink.calcite.shaded.com.google.common.collect.Lists;
import org.apache.flink.configuration.Configuration;
import org.apache.flink.streaming.api.TimeCharacteristic;
import org.apache.flink.streaming.api.datastream.DataStreamSource;
import org.apache.flink.streaming.api.datastream.SingleOutputStreamOperator;
import org.apache.flink.streaming.api.environment.StreamExecutionEnvironment;
import org.apache.flink.streaming.api.functions.KeyedProcessFunction;
import org.apache.flink.streaming.api.functions.timestamps.BoundedOutOfOrdernessTimestampExtractor;
import org.apache.flink.streaming.api.windowing.time.Time;
import org.apache.flink.util.Collector;
import java.util.ArrayList;

/**
 * 恶意登录监控
 */
public class LoginFailTask {
    public static void main(String[] args) throws Exception {
        // 1.创建 Flink 运行环境
        StreamExecutionEnvironment env = StreamExecutionEnvironment.getExecutionEnvironment();
        env.setParallelism(1);
```

第 13 章
Flink 实现电商用户行为分析

```java
        // 2.读取数据
        DataStreamSource<String> inputStream = env.readTextFile(
                "D:\\jareny\\bigdata\\jareny-bigdata-flink-user-behavior-analysis\\src\\main\\resources\\LoginLog.csv");

        // 3.将数据转换为 LoginEvent 用户登录事件实体类对象,并设置事件时间
        SingleOutputStreamOperator<LoginEvent> dataStream = inputStream.map(line -> {
            String[] field = line.split(",");
            return new LoginEvent(new Long(field[0]), field[1], field[2], new Long(field[3]));
        })
                .assignTimestampsAndWatermarks(new BoundedOutOfOrdernessTimestampExtractor<LoginEvent>(Time.seconds(2)) {
                    @Override
                    public long extractTimestamp(LoginEvent element) {
                        return element.getTimestamp() * 1000L;
                    }
                });

        // 4.根据用户 id 分组,判断用户是否需要登录警告
        SingleOutputStreamOperator<LoginFailWarning> result = dataStream
                .keyBy(LoginEvent::getUserId)
                .process(new MyLogin());

        result.print();

        // 5.开始提交到 Flink 执行
        env.execute();
    }

    public static class MyLogin extends KeyedProcessFunction<Long,LoginEvent,LoginFailWarning>{
        // 定义列表状态,保存 2 秒内登录失败的事件
        ListState<LoginEvent> failList;

        // 定义状态:保存注册的定时器时间戳
        // 当有需要删除定时器时,需要一个状态保证注册时的时间戳
        ValueState<Long> timeState;

        @Override
        public void onTimer(long timestamp, KeyedProcessFunction<Long, LoginEvent,
                LoginFailWarning>.OnTimerContext ctx, Collector<LoginFailWarning> out) throws Exception {
            // 定时器触发。说明 2 秒内没有登录成功,判断 failList 中的个数是否大于 2
            ArrayList<LoginEvent> loginFails = Lists.newArrayList(failList.get().iterator());
            int failtimes = loginFails.size();
```

```java
            if (failtimes >= 2){
                // 若大于2,需要输出报警
                out.collect(new LoginFailWarning(ctx.getCurrentKey()
                        ,loginFails.get(0).getTimestamp()
                        ,loginFails.get(failtimes-1).getTimestamp()
                        ,"fail times"+failtimes));
            }

            failList.clear();
            timeState.clear();
        }

    @Override
    public void open(Configuration parameters) throws Exception {
        failList = getRuntimeContext().getListState(new ListStateDescriptor<LoginEvent>("fail list",LoginEvent.class));
            timeState = getRuntimeContext().getState(new ValueStateDescriptor<Long>("time",Long.class));
        }

    @Override
    public void processElement(LoginEvent value, KeyedProcessFunction<Long, LoginEvent,
            LoginFailWarning>.Context ctx, Collector<LoginFailWarning> out) throws Exception {
        // 判断是否登录失败,若登录失败,将事件添加到failList中
        if ("fail".equals(value.getLoginState())){
            failList.add(value);
            if (null == timeState.value()){
                // 若还没有注册定时任务,定一个2秒后的定时任务,更新保存时间戳的状态
                long ts = value.getTimestamp() * 1000 + 2000L;
                timeState.update(ts);
                ctx.timerService().registerEventTimeTimer(ts);
            }
        }else {
            // 若登录成功了,检验一下是否注册了定时器,若注册了,需要删除定时任务
            // 随后清空状态
            if (timeState.value() !=null){
                ctx.timerService().deleteEventTimeTimer(timeState.value());
            }
            timeState.clear();
            failList.clear();
        }
    }
}
```

第 13 章
Flink 实现电商用户行为分析

恶意登记监控使用复杂事件处理（CEP）实现。恶意登记监控使用 Apache Flink 的 CEP 实现是指利用 Flink 的 CEP 库对恶意用户在电商平台上恶意注册的行为进行监控和预警。通过该实现，可以有效地识别出短时间内大量注册、同一设备多次注册等恶意行为，及时触发告警并采取相应的处理措施，保护企业的营销资金和正常用户的权益。

Apache Flink 是一个高性能、高吞吐量的数据流处理框架，提供了事件驱动型的数据流编程模型。CEP 是 Flink 的一个库，提供了复杂事件处理的功能，可以用于识别和匹配一系列事件模式。

```java
package com.jareny.flink.userbehavior.analysis.task;

import com.jareny.flink.userbehavior.analysis.entity.LoginEvent;
import com.jareny.flink.userbehavior.analysis.entity.LoginFailWarning;
import org.apache.flink.cep.CEP;
import org.apache.flink.cep.PatternSelectFunction;
import org.apache.flink.cep.PatternStream;
import org.apache.flink.cep.pattern.Pattern;
import org.apache.flink.cep.pattern.conditions.SimpleCondition;
import org.apache.flink.streaming.api.datastream.DataStreamSource;
import org.apache.flink.streaming.api.datastream.SingleOutputStreamOperator;
import org.apache.flink.streaming.api.environment.StreamExecutionEnvironment;
import org.apache.flink.streaming.api.functions.timestamps.AscendingTimestampExtractor;
import org.apache.flink.streaming.api.windowing.time.Time;
import java.util.List;
import java.util.Map;

/**
 * 恶意登录监控,使用CEP来实现
 */
public class LoginFailWithCep {
    public static void main(String[] args) throws Exception {
        // 1.创建Flink运行环境
        StreamExecutionEnvironment env = StreamExecutionEnvironment.getExecutionEnvironment();
        env.setParallelism(1);

        // 2.读取数据
        DataStreamSource<String> inputStream = env.readTextFile(
                "D:\\jareny\\bigdata\\jareny-bigdata-flink-user-behavior-analysis\\src\\main\\resources\\LoginLog.csv");

        // 3.将数据转换为LoginEvent用户登录行为实体类对象,并设置事件时间
        SingleOutputStreamOperator<LoginEvent> dataStream = inputStream.map(line -> {
            String[] fields = line.split(",");
            return new LoginEvent(new Long(fields[0]), fields[1], fields[2], new Long(fields[3]));
        })
                .assignTimestampsAndWatermarks(new AscendingTimestampExtractor<LoginEvent>() {
                    @Override
                    public long extractAscendingTimestamp(LoginEvent element) {
```

```java
            return element.getTimestamp() * 1000L;
        }
    });

// 4.定义一个匹配模式,用于筛选2秒内都登录失败的事件
    Pattern<LoginEvent, LoginEvent> loginFailPattern = Pattern.<LoginEvent>begin("firstFail").where(new SimpleCondition<LoginEvent>() {
        @Override
        public boolean filter(LoginEvent loginEvent) throws Exception {
            return "fail".equals(loginEvent.getLoginState());
        }
    })
        .next("secondFail").where(new SimpleCondition<LoginEvent>() {
            @Override
            public boolean filter(LoginEvent loginEvent) throws Exception {
                return "fail".equals(loginEvent.getLoginState());
            }
        })
        .within(Time.seconds(2));

// 5.将匹配模式应用到数据流上,得到一个PatternStream
    PatternStream<LoginEvent> patternStream = CEP.pattern(dataStream.keyBy(LoginEvent::getUserId), loginFailPattern);

// 6.检出符合匹配条件的复杂事件,进行转换处理,得到报警信息
    SingleOutputStreamOperator<LoginFailWarning> result = patternStream.select(new LoginFailMatch());

    result.print();

// 7.开始提交到Flink执行
    env.execute();
}

// 实现PatternSelectFunction接口,输出事件
public static class LoginFailMatch implements PatternSelectFunction<LoginEvent, LoginFailWarning> {
    @Override
    public LoginFailWarning select(Map<String, List<LoginEvent>> map) throws Exception {
        LoginEvent firstFail = map.get("firstFail").iterator().next();
        LoginEvent secondFail = map.get("secondFail").get(0);
        return new LoginFailWarning(firstFail.getUserId(), firstFail.getTimestamp(), secondFail.getTimestamp(), "login fail 2 times");
    }
}
}
```

13.7 订单支付实时监控

13.7.1 订单支付实时监控模块介绍

订单失效监控与风险防控是一个基于 Apache Flink 的实时分析系统,用于提高用户支付意愿并降低系统风险。当用户下单后,该模块利用 CEP 库进行事件流的模式匹配,并设定匹配的时间间隔。如果用户在 15 分钟内未支付,则输出监控信息。此外,还可以利用状态编程,通过 Process Function 实现处理逻辑。

13.7.2 订单支付实时监控模块实现

1. 项目目标

对用户订单进行实时监控,以判断是否出现失效事件。当用户下单后 15 分钟内未支付时,触发监控信息输出。通过模式匹配和时间间隔设定,有效识别订单失效事件。降低系统风险,提高用户支付意愿。输出监控信息,提醒管理员及时处理失效订单。

2. 技术方案

使用 Apache Flink 作为流处理框架,实现数据的实时分析和处理。利用 Java 语言编写对象,用于存储和处理数据。利用 CEP 库进行事件流的模式匹配,并设定匹配的时间间隔。利用状态编程,通过 Process Function 实现处理逻辑。使用 Flink 的并行处理能力,对数据进行分布式处理,提高计算效率。

3. 代码实现

1)创建订单事件对象。创建订单事件对象是指用户在电商平台上创建订单操作的相关信息。通过该对象,可以了解用户的购买行为、订单信息以及交易数据等,进一步分析用户的消费习惯、购买偏好和交易趋势等。

```java
package com.jareny.flink.userbehavior.analysis.entity;

import lombok.AllArgsConstructor;
import lombok.Data;
import lombok.NoArgsConstructor;

@Data
@AllArgsConstructor
@NoArgsConstructor
public class ApacheLogEvent {
    // 用户 IP
    private String ip;
    // 用户 ID
    private String userId;
```

```
    // 请求时间戳
    private Long timestamp;
    // 请求方法
    private String method;
    // 请求路径
    private String url;
}
```

2)创建订单状态对象。创建订单状态对象是用于记录和追踪用户在电商平台上下达订单时的状态信息。通过该对象,可以获取订单的审核状态、支付状态、发货状态、收货状态等关键信息,进而深入分析用户的购买行为和交易过程。

```
package com.jareny.flink.userbehavior.analysis.entity;

import lombok.AllArgsConstructor;
import lombok.Data;
import lombok.NoArgsConstructor;

// 订单状态模式
@Data
@AllArgsConstructor
@NoArgsConstructor
public class OrderResult {
    // 订单 ID
    private Long orderId;
    // 订单状态
    private String resultState;
}
```

3)创建收款事件对象。创建收款事件对象是用于记录和追踪用户在电商平台上下达付款指令的重要事件。通过该对象,可以获取付款金额、付款方式、收款方等关键信息,进而深入分析用户的支付行为和交易过程。

```
package com.jareny.flink.userbehavior.analysis.entity;

import lombok.AllArgsConstructor;
import lombok.Data;
import lombok.NoArgsConstructor;

// 订单实时支付
@Data
@AllArgsConstructor
@NoArgsConstructor
public class ReceiptEvent {
    // 支付 ID
    private String txId;
    // 支付渠道
```

```
    private String payChannel;
    // 时间戳
    private Long timestamp;
}
```

4）订单实时监控。订单实时监控是用于实时追踪和监控用户订单状态的重要模块。通过该功能，可以及时发现异常订单、处理订单问题，并为用户提供实时的订单状态更新。

```
package com.jareny.flink.userbehavior.analysis.task;

import com.jareny.flink.userbehavior.analysis.entity.OrderEvent;
import com.jareny.flink.userbehavior.analysis.entity.OrderResult;
import org.apache.flink.cep.CEP;
import org.apache.flink.cep.PatternSelectFunction;
import org.apache.flink.cep.PatternStream;
import org.apache.flink.cep.PatternTimeoutFunction;
import org.apache.flink.cep.pattern.Pattern;
import org.apache.flink.cep.pattern.conditions.SimpleCondition;
import org.apache.flink.streaming.api.datastream.DataStreamSource;
import org.apache.flink.streaming.api.datastream.SingleOutputStreamOperator;
import org.apache.flink.streaming.api.environment.StreamExecutionEnvironment;
import org.apache.flink.streaming.api.functions.timestamps.AscendingTimestampExtractor;
import org.apache.flink.streaming.api.windowing.time.Time;
import org.apache.flink.util.OutputTag;

import java.util.List;
import java.util.Map;

/**
 *  订单实时监控
 */
public class OrderPayTimeoutTask {
    public static void main(String[] args) throws Exception {
        // 1.创建 Flink 运行环境
        StreamExecutionEnvironment env = StreamExecutionEnvironment.getExecutionEnvironment();
        env.setParallelism(1);

        // 2.读取数据
        DataStreamSource<String> inputStream = env.readTextFile(
                "D:\\jareny\\bigdata\\jareny-bigdata-flink-user-behavior-analysis\\src\\main\\resources\\OrderLog.csv");

        // 3.将数据转换为 OrderEvent 订单事件实体类对象,并设置事件时间
        SingleOutputStreamOperator<OrderEvent> dataStream = inputStream.map(line -> {
            String[] field = line.split(",");
            return new OrderEvent(new Long(field[0]), field[1], field[2], new Long(field[3]));
```

```java
}).assignTimestampsAndWatermarks(new AscendingTimestampExtractor<OrderEvent>() {
    @Override
    public long extractAscendingTimestamp(OrderEvent element) {
        return element.getTimestamp() * 1000L;
    }
});

//4.创建一个匹配规则,时间限制的模式
Pattern<OrderEvent, OrderEvent> orderPayPattern = Pattern.<OrderEvent>begin("create")
    .where(new SimpleCondition<OrderEvent>() {
        @Override
        public boolean filter(OrderEvent orderEvent) throws Exception {
            return "create".equals(orderEvent.getEventType());
        }
    })
    .followedBy("pay").where(new SimpleCondition<OrderEvent>() {
        @Override
        public boolean filter(OrderEvent orderEvent) throws Exception {
            return "pay".equals(orderEvent.getEventType());
        }
    })
    .within(Time.minutes(5));

//5.定义一个侧输出流,表示超时事件
OutputTag<OrderResult> outputTag = new OutputTag<OrderResult>("order-timeout"){};

//6.将 pattern 应用到输入数据上,得到 pattern stream
 PatternStream<OrderEvent> patternStream = CEP.pattern(dataStream.keyBy(Order-
Event::getOrderId), orderPayPattern);

// 7.调用 select 方法,实现对匹配复杂事件和超时复杂事件的提取和处理
/**
 * select 方法有三个参数
 * 第一个:侧输出接收对象
 * 第二个:时间超出之后需要处理的方法
 * 第三个:正常事件处理的方法
 */
SingleOutputStreamOperator<OrderResult> result = patternStream.select(outputTag,
new OrderTimeoutSelect(), new OrderPaySelect());

result.print();
result.getSideOutput(outputTag).print("timeout");

//8.开始提交到 Flink 执行
env.execute();
```

第 13 章 Flink 实现电商用户行为分析

```java
    }
    // 实现自定义的超时事件处理函数
    public static class OrderTimeoutSelect implements PatternTimeoutFunction<OrderEvent,OrderResult>{
        @Override
        public OrderResult timeout(Map<String, List<OrderEvent>> map, long l) throws Exception {
            Long timeoutOrserId = map.get("create").iterator().next().getOrderId();
            return new OrderResult(timeoutOrserId,"timeout " + l);
        }
    }

    // 实现自定义的正常匹配事件处理函数
    public static class OrderPaySelect implements PatternSelectFunction<OrderEvent,OrderResult>{
        @Override
        public OrderResult select(Map<String, List<OrderEvent>> map) throws Exception {
            return new OrderResult (map.get ("pay").iterator ().next ().getOrderId (),"payed");
        }
    }
}
```

13.8 订单支付实时对账

▶▶ 13.8.1 订单支付实时对账模块介绍

实时对账与支付信息匹配是一个基于 Apache Flink 的实时分析系统，用于对用户订单的支付信息与到账信息进行匹配核对。该模块从两条流中分别读取订单支付信息和到账信息，通过 connect 连接合并两条流，并使用 coProcessFunction 进行匹配处理。如果发现不匹配的支付信息或到账信息，则输出提示信息，以进行实时对账和纠错处理。

▶▶ 13.8.2 订单支付实时对账模块实现

1. 项目目标

实时读取并分析订单支付信息和到账信息。通过 connect 连接合并两条流，实现信息的匹配核对。使用 coProcessFunction 处理合并流中的数据，进行匹配规则的判断。发现不匹配的支付信息或到账信息时，输出提示信息。支持实时对账功能，提高财务处理的准确性和及时性。

2. 技术方案

使用 Apache Flink 作为流处理框架，实现数据的实时分析和处理。利用 Java 语言编写对象，用于存储和处理数据。利用 Flink 的 connect 方法连接两条流，实现信息的匹配核对。使用 coProcessFunction 作为处理函数，对合并流中的数据进行匹配规则的判断和处理。使用 Flink 的并行处理能力，对数据进行分布式处理，提高计算效率。

3. 代码实现

订单支付实时对账功能是确保订单支付数据准确性和一致性的关键环节。通过实时对账，可以发现并解决支付过程中可能出现的各种问题，如支付金额错误、支付状态不一致等。电商用户行为分析系统中的订单支付实时对账功能的代码实现如下。

```java
package com.jareny.flink.userbehavior.analysis.task;

import com.jareny.flink.userbehavior.analysis.entity.OrderEvent;
import com.jareny.flink.userbehavior.analysis.entity.ReceiptEvent;
import org.apache.flink.api.common.state.ValueState;
import org.apache.flink.api.common.state.ValueStateDescriptor;
import org.apache.flink.api.java.tuple.Tuple2;
import org.apache.flink.configuration.Configuration;
import org.apache.flink.streaming.api.datastream.DataStreamSource;
import org.apache.flink.streaming.api.datastream.SingleOutputStreamOperator;
import org.apache.flink.streaming.api.environment.StreamExecutionEnvironment;
import org.apache.flink.streaming.api.functions.co.CoProcessFunction;
import org.apache.flink.streaming.api.functions.timestamps.BoundedOutOfOrdernessTimestampExtractor;
import org.apache.flink.streaming.api.windowing.time.Time;
import org.apache.flink.util.Collector;
import org.apache.flink.util.OutputTag;

/**
 * 订单支付实时对账
 */
public class TxPayMatchTask {

    private final static OutputTag<OrderEvent> unmatchedPays = new OutputTag<OrderEvent>("unmatched-pays") {};
    private final static OutputTag<ReceiptEvent> unmatchedReceipts = new OutputTag<ReceiptEvent>("unmatched-receipts") {};

    public static void main(String[] args) throws Exception {
        //1.创建 Flink 运行环境
        StreamExecutionEnvironment env = StreamExecutionEnvironment.getExecutionEnvironment();
        env.setParallelism(1);

        //2.读取数据
        DataStreamSource<String> inputStream1 = env.readTextFile(
                "D:\\jareny\\bigdata\\jareny-bigdata-flink-user-behavior-analysis\\src\\main\\resources\\OrderLog.csv");
```

```java
// 3.将数据转换为 OrderEvent 订单事件实体类对象,并设置事件时间
SingleOutputStreamOperator<OrderEvent> orderStream = inputStream1.map(line -> {
    String[] fields = line.split(",");
    return new OrderEvent(new Long(fields[0]), fields[1], fields[2], new Long(fields[3]));
}).assignTimestampsAndWatermarks(new BoundedOutOfOrdernessTimestampExtractor<OrderEvent>(Time.milliseconds(200)) {
            @Override
            public long extractTimestamp(OrderEvent element) {
                return element.getTimestamp() * 1000L;
            }
        });

// 4.读取数据
DataStreamSource<String> inputStream2 = env.readTextFile(
        "D:\\jareny\\bigdata\\jareny-bigdata-flink-user-behavior-analysis\\src\\main\\resources\\ReceiptLog.csv");

// 5.将数据转换为 ReceiptEvent 订单支付实体类对象,并设置事件时间
SingleOutputStreamOperator<ReceiptEvent> receiptStream = inputStream2.map(line -> {
    String[] fields = line.split(",");
    return new ReceiptEvent(fields[0], fields[1], new Long(fields[2]));
}).assignTimestampsAndWatermarks(new BoundedOutOfOrdernessTimestampExtractor<ReceiptEvent>(Time.milliseconds(200)) {
            @Override
            public long extractTimestamp(ReceiptEvent element) {
                return element.getTimestamp() * 1000L;
            }
        });

// 6.将订单 id 为空的数据过滤,之后使用 connect 连接数据
SingleOutputStreamOperator<Tuple2<OrderEvent, ReceiptEvent>> result = orderStream
        .filter(data -> !"".equals(data.getTxId())).keyBy(OrderEvent::getTxId)
        .connect(receiptStream.keyBy(ReceiptEvent::getTxId))
        .process(new TxPayMatchDetect());

result.print("normal");
result.getSideOutput(unmatchedPays).print("unmatchedPays");
result.getSideOutput(unmatchedReceipts).print("unmatchedReceipts");

// 7.开始提交到 Flink 执行
env.execute();
}

// 继承 CoProcessFunction 类,用于处理连接的数据
public static class TxPayMatchDetect extends CoProcessFunction<OrderEvent,ReceiptEvent,Tuple2<OrderEvent,ReceiptEvent>>{
```

```java
    // 设置两个状态,保存连接的数据
    ValueState<OrderEvent> payState;
    ValueState<ReceiptEvent> receiptState;

    @Override
    public void onTimer(long timestamp, CoProcessFunction<OrderEvent, ReceiptEvent,
            Tuple2<OrderEvent, ReceiptEvent>>.OnTimerContext ctx, Collector<Tuple2<OrderEvent, ReceiptEvent>> out) throws Exception {
        // 判断状态中是否有值,而不是状态本身是否为 null
        // 定时器触发,有可能是有一个事件没来,不匹配,也有可能是都来过了,已经输出并清空状态
        // 判断哪个不为空,那么另一个就没来
        if (payState.value() != null){
            ctx.output(unmatchedPays,payState.value());
        }

        if (receiptState.value() != null){
            ctx.output(unmatchedReceipts,receiptState.value());
        }

        payState.clear();
        receiptState.clear();
    }

    @Override
    public void open(Configuration parameters) throws Exception {
        payState = getRuntimeContext().getState(new ValueStateDescriptor<OrderEvent>("pay state",OrderEvent.class));
        receiptState = getRuntimeContext().getState(new ValueStateDescriptor<ReceiptEvent>("receipt state",ReceiptEvent.class));
    }

    @Override
     public void processElement1(OrderEvent value, CoProcessFunction<OrderEvent, ReceiptEvent,
            Tuple2 < OrderEvent, ReceiptEvent > >. Context ctx, Collector < Tuple2 < OrderEvent, ReceiptEvent>> out) throws Exception {
        // 订单支付事件来了,判断是否已经有对应的到账事件
        ReceiptEvent receiptEvent = receiptState.value();
        if (null != receiptEvent){
            // 如果 receipt 不为空,说明到账事件已经来过,输出匹配事件,清空状态
            out.collect(new Tuple2<>(value,receiptEvent));

            payState.clear();
            receiptState.clear();
        }else {
            // 如果 receipt 没来,注册一个定时器,开始等待,并更新状态
```

```java
            ctx.timerService().registerEventTimeTimer((value.getTimestamp() + 5) * 1000L);
            payState.update(value);
        }
    }

    @Override
    public void processElement2(ReceiptEvent value, CoProcessFunction<OrderEvent, ReceiptEvent,
            Tuple2<OrderEvent, ReceiptEvent>>.Context ctx, Collector<Tuple2<OrderEvent, ReceiptEvent>> out) throws Exception {
        // // 到账事件来了,判断是否已经有对应的订单事件
        OrderEvent orderEvent = payState.value();

        if (null != orderEvent){
            // 如果 order 不为空,说明订单事件已经来过,输出匹配事件,清空状态
            out.collect(new Tuple2<>(orderEvent,value));

            payState.clear();
            receiptState.clear();
        }else {
            // 如果 order 没来,注册一个定时器,开始等待,并更新状态
            ctx.timerService().registerEventTimeTimer((value.getTimestamp() + 3) * 1000L);
            receiptState.update(value);
        }
    }
}
```